Engineering Peace and Justice

P. Aarne Vesilind

Engineering Peace and Justice

The Responsibility of Engineers to Society

Springer

P. Aarne Vesilind, PhD
Bucknell University
Dept. Civil & Environmental Engineering
Lewisburg PA 17837
USA
vesilind@bucknell.edu

ISBN 978-1-4471-5822-6 ISBN 978-1-84882-674-8 (eBook)
DOI 10.1007/978-1-84882-674-8
Springer London Dordrecht Heidelberg New York

British Library Cataloguing in Publication Data
A catalogue record for this book is available from the British Library

Cover design: eStudioCalamar, Figueres/Berlin

Printed on acid-free paper

Springer is part of Springer Science+Business Media (www.springer.com)

Preface

Some years ago when I was chair of the department of civil and environmental engineering, a colleague introduced me to a visitor from Sandia Laboratories, perhaps the largest developer of armaments and weapons systems in the world. We had a nice visit, and as we chatted, the talk naturally centered on the visitor's engineering work. It turned out that his job in recent years had been to develop a new acoustic triggering device for bombs. As he explained it, the problem with bombs was that the plunger triggering mechanism could fail if the bomb hit at an angle, and thus the explosives would not detonate. To get around this, he developed an acoustic trigger that would detonate the explosives as soon as the bomb hit any solid surface, even at an angle.

As he talked, I watched his face. His enthusiasm for his work was clearly evident, and his animated explanations of what they had developed at Sandia exuded pride and excitement.

I thought about asking him what it felt like to have spent his engineering career designing better ways to kill people or to destroy property – the sole purpose of a bomb. I wondered how many people had been killed because this man had developed a clever acoustic triggering device.

But good sense and decorum prevailed and I did not ask him such questions. We parted as friends and in good spirits.

Afterwards I thought about him as an exemplar of an engineer who is so focused on engineering that it would never occur to him to ask himself what his work is good for and how it might be used. Many engineers are like that – they put on blinders when it comes to their work so that they don't see the moral implications.

This book is about, and mainly for, engineers who have concerns about their own work and the profession of engineering. My argument is that engineering can in a rough way be divided into three categories: military engineering, civilian engineering, and what I call peace engineering. I demonstrate the essence of these divisions within professional engineering by using short biographies of real engineers who have, in my opinion, exemplified these three engineering paradigms. My hope is that, in reading about exemplary engineers, others will be able to better understand their own jobs and, most importantly, recognize that there are alternative engineering careers available to them that would proactively promote peace and justice for all.

Much material in this book is the result of a long career in engineering education and practice, and I am indebted to the many engineers with whom I have had discussions on the role of professional engineering in our society. The concept of "peace engineering" was formulated when I was the R. L. Rooke Chair of the Historical and Societal Context of Engineering at Bucknell University, and I am grateful to the faculty and students at Bucknell for their support and encouragement.

The entire manuscript was read by Pamela Vesilind and Libby Vesilind, both of whom are not only fine writers but also effective critics who did not hesitate to challenge me on elements of grammar, style, or logic. Their hard work is very much appreciated.

August Rebane (1885–1965)

Finally, I want to acknowledge the influence my maternal grandfather, August Rebane, had in defining my own world view, or *Weltanschauung*. When I last saw my grandfather I was 5 years old and my family was fleeing Estonia to escape the Soviet Red Terror in 1944. So I cannot say that I learned anything from him in the normal way a grandfather teaches a grandson. But I know him through my mother's stories and recollections. And even many years after his death, I feel his presence in my own thoughts and beliefs. He was an idealist who sought freedom for Estonia, and then did his best to establish a socialist government in the newly formed republic, believing that this was the fair and equitable thing to do. He was a gentle pacifist whose friends loved him for his humor and kindness. I know who I am because of the bond of kinship between my grandfather and me, and it is therefore with humility and gratitude that I dedicate this book to August Rebane.

New London, NH, 2010 *P. Aarne Vesilind*

Contents

Chapter 1
The Evolution of the Engineering Profession

Engineering has a long and proud history. Ever since humans began to live in settlements, the need for those who knew how to get things done was critical to survival. It is not difficult to imagine how the earliest "engineers" invented the wheel, the plow, and the bow and arrow. As cities and civilizations developed, the value of engineers to those in power steadily increased.

The fact that the name "engineer" has survived many revolutions and has entered into use in many languages indicates that engineers must have been central to all human political organizations. Bert Young, a linguist, describes the lineage of *engineer* in a 1914 article published by the American Society of Civil Engineering in their *Transactions*. According to Young, the word *engineer* comes from the Middle English *engyneour* and from the Old French *engignier* or *engigneour*. These in turn come from Middle Latin *ingeniarius*, defined as one who makes or uses an engine, especially a war engine. *Engine* is from Latin *ingenium*, an invention, an engine. Young observes that "there must have been confusion of Latin *ingenuus* and Latin *ingeniosus*. These should be almost opposite in meaning. I suppose an *engineer* ought to be both *ingenious* and *ingenuous*, artful and artless, sophisticated and unsophisticated, bond and free" [1].

The greatest job satisfaction for engineers is watching something they conceive, design, and construct actually perform as intended. Samuel Florman beautifully describes this joy as an "existential pleasure" – existential in that the engineer is free of concerns for how the results of engineering will be used. How a product of engineering is used does not matter, argues Florman, as long as it works as intended. This gives the engineer the existential freedom to do good engineering and not be concerned about what the product or facility will eventually be used for or who will use it [2].

The danger and challenge of engineering is that not every clever idea works as it is supposed to. Sometimes the best intentions and best technical skills are not enough, and what seems like a good idea turns out to be disastrous.

- The bridge built in 1940 over the Tacoma Narrows in Washington State was as spectacular structurally as it was beautiful to the eye. It seemed to float across the water. But soon after completion, people began to notice that the bridge would sway in windstorms. It turned out that in trying to make the bridge light and beautiful, the design engineers had constructed the decking like an aircraft wing, which behaved as such under windy conditions. Because the decking would bounce and sway in windstorms, the bridge became affectionately known as the "Galloping Gertie." Then on 7 November 1940 a strong wind caused the bridge to collapse catastrophically (Fig. 1.1) [2].
- Sometimes the very cleverness of an engineering solution can create other problems. In Bangladesh, the World Health Organization responded to the problem of arsenic in the drinking water by installing thousands of ion exchange resin canisters (much like the activated carbon canisters used in some homes to improve the taste of drinking water) that absorbed arsenic ions. The system worked well, until the villagers asked what they should do with the used canisters, which then contained a high concentration of arsenic and were clearly hazardous waste that needed careful handling and disposal. Unfortunately, the WHO engineers had not thought this through; now Bangladesh has tens of thousands of these canisters that may eventually cause acute human health problems [3] (Fig. 1.2).
- Some engineering projects are so blatantly wrong that one wonders why they were ever undertaken. The classic case of the US Army Corps of Engineers project for draining the Everglades is typical of this category of engineering work. Wanting to increase arable land for development and farming, the Corps constructed miles of waterways that drained the groundwater from the swamp, at great cost to the American taxpayer. However, this lowered the groundwater level, dried up the swamps, and gave rise to huge forest fires in the Everglades.

Fig. 1.1 The collapse of the first Tacoma Narrows bridge

The most devastating effect was the loss of wildlife that depended on swampy conditions. To their credit, the Corps finally realized their mistake and spent many more millions of tax dollars to correct the mistakes they had made. Well-intentioned engineering, such as draining the Everglades, can have negative unintentional consequences that can lead to other problems [4] (Fig. 1.3).

As the above examples illustrate, engineering, done with the best intentions, can result in catastrophes. This is also true for the other professions. Some actions by professionals such as physicians and lawyers can likewise have unintended and negative consequences. For example, a physician might treat a patient for a certain symptom, but in so doing create other unintended health problems. Such physician-caused health problems are called *iatrogenic* diseases.

Drew Endy, an engineer at Stanford University, has proposed a new word that describes unintentional problems created by engineering. He calls them *mechanikogenic* problems, the unintended detrimental outcomes of engineering. Just as the US Army Corps of Engineers had good intentions in draining the Everglades, their actions unfortunately caused great harm that could be described as mechanikogenic damage to the Everglades ecology.

One mechanikogenic effect of engineering technology is making possible the waging of modern warfare. Engineers who find joy in solving problems may discover that their efforts have made possible the waging of war and the killing of countless innocent people. This is an unintended collateral effect of engineering, and warfare becomes a mechanikogenic problem.

Since ancient times, engineers have been essential in the construction of defenses for human settlements such as villages and cities. Some engineers, on the other hand, have been employed by their rulers to design the machines that will defeat the defenders of human settlements. Engineers whose primary occupation is either in national defense or in the design and use of offensive weapons are called *military engineers.* On the other hand, those engineers who don't do military work and instead practice their craft to benefit people and to improve their quality of life are known as *civilian engineers.* These two broad categories of engineering reach far back into history.

Fig. 1.2 Arsenic removal system in Bangladesh [courtesy of Arup SenGupta]

Fig. 1.3 Drainage of the Everglades [courtesy of Florida State Archives]

1.1 Early Military Engineering

War has always been a part of the human condition, and engineering, the application of science to produce useful things, has always been essential to the waging of war. The importance of engineering (or at least the discovery that tools can be used to kill others) is famously illustrated in a scene from Stanley Kubrick's classic film *2001: A Space Odyssey* (Fig. 1.4). One of our ancestors, surveying a pile of bones, slowly begins to realize that if he picks up a bone and swings it around, he can use it to hurt others. In Kubrick's film, these primitive clubs became the first implements designed (by the engineer?) as tools of war.

Ancient civilizations valued engineers. The Romans had their engineers travel with the military legions, and the engineers were responsible for constructing fortifications and base camps, using local materials and requisitioned labor. Engineers were a special class of soldier, free from regular duties, and they could call on regular soldiers to assist in the construction of camps and forts that were built according to preset plans. The complexity of the structures depended on how long the camp was to be used. Roman engineers apparently could erect legion camps in a single night. One of their greatest feats was the construction of a ditch across the

Fig. 1.4 Scene from Stanley Kubrick's film *2001: A Space Odyssey* [Courtesy of Metro-Goldwyn-Mayer]

heel of Italy in order to contain Hannibal's armies [5]. Another stupendous effort was the construction of a 13-mile-long double wall around the city of Alesia. The wall prevented attacks from both the city and reinforcements seeking to relieve the besieged city. The presence of the wall made it possible to starve out the defenders and to capture the city [6].

Roman engineers were especially adept at building bridges, many of which survive today. They figured out how to use keystone and arch construction for long spans such as the 100-m-long bridge over the Rhine River built by Julius Caesar's engineers. The Romans also excelled in the construction of siege machines, many of which were adaptations from earlier Greek designs. Road construction was a large part of the Roman military agenda because a road system allowed them to supply both reinforcements and supplies to remote areas.

The point of warfare is to kill the opponent, and any means, including chemical warfare, has been employed since ancient times. The earliest known example of chemical warfare was unearthed at an archaeological site in the Syrian Desert. Soldiers of an ancient Persian empire apparently gassed a platoon of Roman troops, asphyxiating them with smoke from burning bitumen and sulfur. A grave of 20 Roman soldiers in full battle armor were discovered at the site of the ancient city of Dura-Europos in the 1930s, but it was only recently that archeologists determined that those soldiers were killed by poison gas while digging tunnels under the walls of the city. In 189 BC, at the siege of Ambracia, Roman troops were probably killed by ammonia gas produced by burning bird feathers [7]. Chemical warfare, the responsibility of chemical engineers, seems to have been more common in ancient times than had been originally thought.

Whether they used chemical or physical weapons, engineers have always been valuable to kings, potentates, and religious leaders who wanted either to wage war or to defend their possessions against enemies. As a result, in ancient times, the men who were called engineers were almost exclusively concerned with war and defense – doing military engineering. As technology developed, both for offensive and defensive purposes, so did the importance of the military engineer. During the Middle Ages, for example, castles were impressive defensive structures, and ingenious devices were developed to attack these castles.

Although bows had been in use for centuries, the catapult could be considered the first actual "engine of war," since for the first time it was possible to *store* energy and then release it at will. Before then, battles were won by the use of tactics and by marshalling men to heights of valor. The energy in a battle was always human energy. The catapult, however, was an inanimate device that was used for storing potential energy and then converting this to kinetic energy. The early catapults were crude affairs and terribly inaccurate, but they continued to be improved. The most devastatingly effective offensive engine was the trebuchet (Fig. 1.5), which used a sling for hurling large rocks against a castle wall. Because the same amount of energy could be stored for each shot (the counterweight on the other end of the swinging arm), trebuchets were amazingly accurate over a range exceeding 100 m. No castle was safe from these devices, and sieges invariably ended with surrender [8].

But engineers charged with designing defenses were not idle. During the Middle Ages, Italian military engineers figured out how to defend against the trebu-

Fig. 1.5 Trebuchet, one of the most effective engines of war during the Middle Ages

chet by building moats and sloped walls that countered the trebuchet's effectiveness (Fig. 1.6). The new forts had bastions reaching out from the low walls and sloping directly into the moats. Defenders on the bastion had a free range of fire sideways, thereby protecting each other. Not to be outdone, the offense introduced a new and devastating weapon – the cannon. Siege of forts now required that cannons be maneuvered close enough to a wall to blast a breach through it, providing a path for assault. Over many hundreds of years the balance between offense and defense was maintained by engineers who continued to develop both new weapons and more effective defenses.

As military engineers continued to improve weaponry, each significant improvement was thought to be so devastating as to prevent future warfare. Of particular interest is the crossbow, which was considered an unfair weapon because it could pierce armor. As a result, its use against Christians was banned by the Lateran Council in 1139, although its use against heathens was allowed and even encouraged [5]. Such prohibitions had as much effect then as prohibitions against the use of weapons of mass destruction have in modern times.

The engineers who designed and built all these devices and fortifications often worked on a contractual basis. They usually did not take sides, but offered their services to those who paid the most. During the First Crusade, for example, a local specialist was given the task of constructing the siege towers that were essential in the fall of the castle at Nicaea. He was reportedly well paid for his efforts [9].

As the complexity of both offensive weapons and defensive structures increased, the value of engineers to their employers also increased. Frederick II was so fond of one of his engineers, Calamandrinus, that he had Calamandrinus chained up so he could not escape to ply his trade elsewhere. Calamandrinus did escape, however, and did exactly that, offering his engineering skills to Frederick's opponents, who, among other rewards, provided him with a wife.

Master Bertram, born in 1225, became a royal engineer for Henry III of England. In an early record, Bertram *le Engynnur* was one of six such men rewarded by the king for his services. By 1276 he was employed in making engines for the

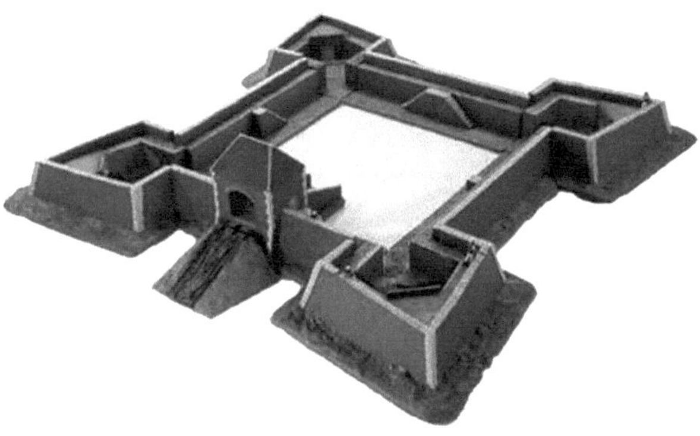

Fig. 1.6 A fort in the Italian style [courtesy of JRMiniatures]

Tower of London, and apparently he took great pride in this work, taking personal charge of buying the oak, beech, and elm from local forests and supervising the construction. Later in his career he was in charge of constructing siege engines in Wales, or Dolwyddelan, and in 1283 at Castell-Y-Bere, where he was referred to as *machinator* and *ingeniator*. Before his death in 1284, Bertram had also built some of the earliest Welsh castles. He was, in short, a talented engineer willing to sell his services to the highest bidder. He built both castles in Wales as well as the siege engines to destroy the very castles he had built [8].

As nation-states became more organized, engineers were less likely to sell their services to the highest bidder and tended to work for only one country, leading to competition among military engineers from different countries. During the 17th and 18th centuries the best military engineers were French. The French engineer B. F. deBelidor (1698–1761) was the real inventor of the shell commonly attributed to the British Major Henry Shrapnel. Pontoon Bridges, designed by F. J. Camus in 1710 and D'Herman in 1773, carried French armies marching three abreast over rivers they encountered. Gun carriages designed by C. F. Berthelot (1718–1800) set a pattern for French artillery. The semaphore telegraph, consisting of a series of towers in visual sight of each other, was invented by French engineer Ignace Tresaguet (1760–1828) for the purpose of providing rapid communication for French armies. Awarded the title of l'Ingénieur Télégraphe, his system connected Paris to Lille, Strasbourg, Brest, and Lyon, and this enabled Napoleon to sustain his conquest of Italy by linking Lyon with Turin, Milan, and Venice [10].

The best engineering school in the world in those years was, not surprisingly, L'Ecole Polytechnique in Paris. The original structure of L'Ecole Polytechnique still stands today, and the frieze on the front of the school signifies the importance of military engineering (Fig. 1.7). French engineering was so influential that the official language of engineering was French, and because many textbooks were in French, the curriculum at the US Military Academy at West Point in the early 19th century included instruction in the French language. Even the motto of the US Army Corps of Engineers is in French: *essayons*, meaning *let us try*.

In the USA the first engineers were often self-taught men with little education, but a lot of imagination. George Washington, for example, taught himself how to survey, and Thomas Jefferson designed and built clever mechanical devices. The Founding Fathers respected engineering and admired the ability to get things done.

Fig. 1.7 Frieze on the front of L'Ecole Polytechnique in Paris

In 1777 George Washington wrote:

> The want of accurate maps has been a grave disadvantage to me. I have in vain endeavored to procure them, and have been obliged to make shift with such sketches as I could trace out of my own observations and that of gentlemen around me. I think if gentlemen of known character and probity could be employed in making maps (from actual surveys) it would be of the greatest advantage [11].

The engineering education that officers received at the US Military Academy at West Point was an obvious advantage on the battlefield during the Civil War – for both sides. After the war, many engineers educated at the military academy went on to do civilian work, and eventually the need for such engineers prompted the founding of new schools to educate nonmilitary, or civilian, engineers.

1.2 Early Civilian Engineering

Civilian engineering stretches back to the beginning of recorded history. Imhotep (Fig. 1.8), who worked for the Egyptian pharaoh Djoser around 2600 BC, is generally acknowledged as the first engineer in history. The name Imhotep means "the one who comes in peace." He was responsible for building the first great pyramid and he probably had a hand in building several others. In so doing, he had to solve many engineering problems, including figuring out how to create a level foundation. Most Egyptologists believe that he did this by digging trenches to the Nile River and allowing the water to flow into the channels in the foundation, thus establishing benchmarks from which to build. He also figured out how to cut the stone and transport the huge blocks to the pyramid. It is still unclear how all this was accomplished, but graves from the time of the construction show many men with shattered bones and spines, attesting to the human cost of construction. Imhotep was able to ply his engineering trade in civilian construction because the kingdom at that time was secure and at peace, and military engineering was not needed.

Fig. 1.8 Detail from a bronze statue of Imhotep

In more modern times, the first formal recognition of civilian engineering, as opposed to military engineering, occurred in England during the 18th century. Because England was protected by surrounding water from large-scale invasion (the only invasion to have ever occurred being in 1066), engineering skills were able to develop more rapidly in the civilian than in the military sector. After the formation of what is now the UK, there was little need for forts or for engines of war, and engineers were able to concentrate on civilian interests.

In the UK, the schools where engineering was taught differed from those in France. Whereas in France engineering education took place in separate schools, in England engineering education developed as parts of existing schools of higher education because there was a need for technology in the local industries. Engineering programs in many of these colleges were quite narrow, reflecting the needs of the industry in that region. For example, Newcastle Royal College of Science specialized in mining; Yorkshire College, Leeds, grew in the textile country, and University College, Liverpool, in a commercial setting; and the Mason Science College, Birmingham, was so vocational that theology and literary study were excluded from its curriculum. Even when this college became a university in 1900, one of the departments was named Commerce and within it was a Department of Brewing [12].

The distinction between military engineering and civilian engineering is admittedly a fuzzy one, and England had its share of military engineers. Similarly, many French engineers did civilian work. But the core of French engineering was military and the majority of English engineers worked on civilian projects.

The distinction between civilian and military engineering is also muddied by the fact that it is often difficult to distinguish whether a project is primarily for civilian use or for use by the military. For example, the construction of lighthouses around the British Isles was important for both naval and commercial vessels. Lighthouses are used by all ships irrespective of their mission. In our own time, the Interstate Highway System, which is now used almost exclusively for civilian purposes, was actually built with a military objective – to be able to move troops rapidly in case the USA was ever invaded. The work done by the transportation engineers who constructed the Interstate Highway System could therefore be classified as civilian engineering, even though the highways might someday also have military uses.

In the USA, as the importance of civilian work increased, universities were established to fulfill the need for such engineers. The first nonmilitary engineering school in the USA was established at Troy, NY, as the Rensselaer Polytechnic Institute, and soon other private engineering schools were founded, including Cooper Union, Stevens, Lehigh, and others. Land grant colleges opened in the 1860s and produced a steady stream of educated American engineers. Because these engineers were engaged in public (civilian) work as opposed to military work, they became known as *civil engineers*. In 1852 American civil engineers formed a professional organization, the American Society of Civil Engineers (ASCE), that was patterned after medieval guilds. Part of the function of the guilds, and ASCE, was to govern the actions of its members and to set standards for entry into the profession.

By the middle of the 19th century engineers had become conscious of the fact that they were different from other people in the construction industry such as architects and carpenters. They realized that what distinguished engineers from those who might also build or manufacture things was their ability to use models to predict the physical world. Engineers were able to estimate fairly accurately whether a building or a bridge would or would not fall down under a given load or whether water in a pipe would flow at the desired velocity. The incorporation of higher mathematical skills, particularly knowledge of calculus, elevated engineering into a profession.

At the time of the founding of ASCE, all professional engineers were builders and equated civil engineering with civilian engineering. In order to become a member of ASCE, an engineer had to be sponsored by ASCE members, and then voted on by the entire membership, not unlike a medieval guild or a modern social fraternity.

In the late 1800s a group of engineers who worked on the manufacture of steam engines, locomotives, and other machines and who thought of themselves as doing civilian (civil) engineering asked to join ASCE, but they were rejected. The old fogies in the society had a very narrow definition of engineering – they believed that you were not a *real* engineer if you did not build structures. The machine builders did not waste much time forming their own professional organization, which became the American Society of Mechanical Engineers (ASME). A few decades later, electrical engineers formed what eventually became the Institute of Electronic and Electrical Engineers (IEEE), the chemical engineers organized the American Institute of Chemical Engineering (AIChE), and so on. What started out as a single profession is today fragmented into subdisciplines, and this process seems to be continuing as new fields develop to meet societal needs.

Even with this fragmentation, engineering has retained its status as a profession. One criterion of a professional is that he or she is licensed by the state to perform functions that are useful to society. Professional engineers who are licensed by the state have special privileges. For example, only licensed professional engineers are allowed to testify as engineering experts in legal cases, and only licensed engineers are allowed to approve designs for construction projects when public funds are involved. Engineers are also allowed self-regulation of the profession, and engineers are still the gatekeepers who determine who may be called an engineer. Because society gives engineers such privileges, the public expects certain benefits from the profession such as honesty, truthfulness, and a commitment to public service. An implied "professional contract" exists between engineers and the public.

This "professional contract" is strikingly similar to the "social contract" theory of Thomas Hobbes. To Hobbes the strongest argument for behaving morally in everyday life is that moral behavior is beneficial to all and that we therefore have a "social contract" or agreement among ourselves to act morally. I suggest that a similar "professional contract" exists between engineers and the public. Although it is not directly reciprocal, there is an understanding of behavior that, when adhered to, will benefit all. Engineers spell out this contract with the public in a code of ethics for the profession.

1.3 The Emergence of Peace Engineering

Up to the present day we have seen two types of engineers – those whose primary work is in military engineering and those whose jobs are mostly in the civilian sector. I would now like to argue that over the past few decades a third kind of engineering career has been defined by men and women who not only seek to serve the public good, but who want to work purposefully for peace and justice. These engineers recognize that military engineering, in all its forms, including working for defense contractors and conducting research for the Department of Defense, is destined to be used for warfare, either defensive or offensive, and they are unsure if they want to participate in such work. Engineers in the civilian sector might question if their skills and talents are being used in the best way possible. Is the design and construction of another big-box superstore, for example, really the best way to spend limited global resources? Some military and civilian engineers are looking for alternatives that would allow them to use their engineering skills in a positive and proactive way to promote peace and justice. These engineers, by their actions, are defining a third kind of engineering – *peace engineering.*

Peace engineering is the use of engineering skills to promote a peaceful and just existence for all people. Examples of how engineers have used and will continue to use their skills for this purpose include engineers who work for the Peace Corps, the World Bank, Pan-American Health Organization, and perhaps hundreds of nongovernmental organizations such as Engineers Without Borders.

Some engineers come to peace engineering after having been, perhaps unknowingly, practicing military engineering. Because the Department of Defense funds large research projects at both private and public universities, and contracts out the construction of weapons systems to corporations, many engineers work as military engineers without realizing the purpose of their work. For example, an acquaintance of mine who worked for General Electric thought he was working on a new technology for toasting bread. Only accidentally did he discover that his engineering calculations were clandestinely being used in the Star Wars project promoted by President Reagan. His keen disappointment with that secrecy and deception led him to resign his job at the company.

As with my friend, engineers who are concerned about how their engineering work is to be used are beginning to ask disquieting questions. For example, they wonder if, by working either directly or indirectly for the military establishment, they are truly living up to their professional code of ethics. Engineering, as a profession, states its purpose and objectives in a code of ethics, and at least in the USA, the code of ethics of almost every engineering discipline begins with the following statement:

> The engineer, in his professional practice, shall hold paramount the health, safety, and welfare of the public.

The two key words are *shall* and *paramount*. There is no equivocating about this as the primary commitment of engineering, and almost all engineers agree with this statement and practice accordingly.

There is, however, a problem with this statement for engineers working in the military, and that problem centers on the word "public." What exactly is the "public"? Suppose an engineer works for a company that designs and produces landmines. Is "public" the people who pay his or her salary? Has the "public" decided, through a democratic process, that the manufacture of landmines is necessary? Or is the "public" of record those people who will eventually have to walk over the ground in which these landmines have been planted and be killed and maimed by the explosions?

Many engineers have begun to realize that their responsibility to society is greater than just doing a competent job as technicians. The very fact that they have the skills to put knowledge to work that can cause either good or harm to others requires that they channel these skills in a moral way. Not everyone agrees with this view, of course. Many people believe that engineers should think of themselves as hired guns, doing the bidding of their clients and employers, and not asking questions about the morality of their work. My belief, shared by the overwhelming majority of the profession, is that we do owe moral consideration to those we serve. The purpose of this book is to ask difficult questions about what the role of engineers in our society should be and to encourage engineers to think about their personal engineering careers. The best outcome for any engineer would be a convergence of career goals and personal values.

References

[1] McDonald H (1737, 1914) Origin of the word "engineer." ASCE Trans 77:1737; reproduced in: ASCE Committee on History and Heritage of American Civil Engineering (1970) The civil engineer: his origins. Historical Publication No. 1

[2] Petroski H (1992) To engineer is human: the role of failure in successful design. Vintage, New York

[3] SenGupta A (2008) Arsenic removal from groundwater and its safe containment in a rural environment: validation of a sustainable approach. Environment Sci Technol 42:4268–4273

[4] Grunwald M (2007) The swamp: the Everglades, Florida, and the politics of paradise. Simon & Shuster, New York

[5] van Creveld M (1989) Technology and war. Free Press, New York

[6] Grant M (1978) The history of Rome. Prentice-Hall, Englewood Cliffs, NJ, p 52

[7] Connors S (2009) The Independent, London, 29 January 2009

[8] Bradbury J (1992) The medieval siege. Boydell, Woodbridge, UK

[9] Armytage WHG (1996) A social history of engineering. Westview, Boulder

[10] Hill D (1984) A history of engineering in classical and medieval times. Open Court, LaSalle, IL

[11] US Army Corps of Engineers Web page: http://www.usace.army.mil/Search/lostpages.aspx

[12] Garrison E (1991) A history of engineering and technology. CRC, Boca Raton, FL

Chapter 2
Military Engineering

The military establishment is ubiquitous and omnipresent in American life. Almost everything is of concern to the military, including housing, medicine, agriculture, manufacturing, and transportation. It is impossible to escape the reach of the military because everything has some military relevance.

Resources devoted to military work are staggering. The USA is responsible for over 40% of the more than $1.2 trillion dollar global military expenditure, with the UK second, accounting for 20% [1]. In the UK, funding of military research topped $4 billion in 2003–2004, with 40% of all government researchers employed by the Ministry of Defense [2]. The single largest employer of engineers in the UK is an armaments company [3].

The reason for the huge expenditure in the USA is that the defense establishment believes that it is necessary to have military forces and technology second to none. In the *1997 Quadrennial Defense Review* the Department of Defense reaffirmed that "it is imperative that the United States maintain its military superiority ..." During that same year, the National Defense Panel reported that if the United States does "not lead the technological revolution we will be vulnerable to it," and in the Senate Armed Services Committee Report, the committee wrote that its priority is "to maintain a strong, stable investment in science and technology in order to develop superior technology that will permit the United States to maintain its current military advantages ... and hedge against technological surprise" [4].

Military work is by far the most important source of employment and income for American engineers. Two thirds of all scientists and engineers in the USA work directly or indirectly on defense contracts, and about 60% of all federally funded research is defense related [5]. Papadoupoulos conducted a study of American engineers in defense-related work and concluded that 8.7% of engineers, compared to 2.9% of all workers (including engineers), work at a company or organization that can be described as defense related.

A disproportionate number of engineers are engaged in military activities. According to the US Bureau of Labor Statistics and the Department of Defense, about 3% of the total work force in the USA is employed in military-related work, while over 10% of the engineers work in defense-related industry [6]. Engineers have the highest fraction of their research and development funding from the Department of Defense, accounting for fully 33% of their total funding [6].

Because of the many technical jobs available in armaments or weapons systems, most engineers have little choice but to work on military projects. In a free market democracy, it should theoretically be possible to move between different types of jobs, but this is not easy for engineers. Most engineers are not privately employed but instead work for employers who must sell their collective engineering skills as an entire company. A consulting engineering firm, for example, obtains business from clients because it convinces the client that the collection of engineers employed by that firm can perform the required job. A single engineer, working by himself or herself, is severely limited in the types of work he or she can do.

Because large engineering projects can be compartmentalized, it is often impossible for one engineer to tell just what the overall objective of the project is. The engineer who thought he was working on a toaster and was instead participating in the Star Wars project is an example. At other times, the result of engineering projects can be put to unintended use. Columbia University electrical engineering professor Steve Unger relates a story where he, while working for Bell Labs, was asked to work on an electronic telephone switching system. He believed that such a telephone system was clearly beneficial to the public, but after working for 2 years on the project, he found that the switching software they developed was of interest to the military and the first use of the compiler was by the military. Without knowing it, he had been doing military engineering (s. Unger, pers. commun., 2006).

One reason so many graduates of engineering schools go to work in military engineering is that, quite honestly, it is exciting work. Peter Singer, author of *Wired for War,* a bestselling book on the use of robots in warfare, told an interviewer, when asked why he wrote a book about robots, "I think the opening line of my book explains it all: 'Because robots are freakin' cool'" [7]. Many projects undertaken by military engineers *are* cool. The robotic, or unmanned, systems are especially neat. You design the mechanism and software, and then you program it to do what you tell it to do. How cool is that? It is almost beside the point that the robot is designed to kill someone, or, if the robot malfunctions, it is capable of killing you without any moral qualms.

One way of illustrating what military engineering is all about is to tell the stories of engineers who have devoted their careers to military matters. There are, of course, hundreds of thousands of stories about engineers who have had military engineering careers. I have selected a few exemplars that illustrate both the best and worst of military engineering, with the objective of defining what is meant by the term "military engineering." The purpose in telling their stories is to illustrate how military engineering is both timeless and ubiquitous. Some stories are about engineers who work directly for the military, while other stories focus on engineers who might have been surprised how their engineering invention or effort ended up being applied for a military purpose.

The use of short biographies in order to define the field is akin to defining a genre of art by visiting a museum and studying the paintings of many artists. A visit to the

Musée d'Orsay in Paris, for example, will reveal a large and varied collection of Impressionist art. Paintings by Monet can be viewed next to works by Cézanne, van Gogh, Gauguin, and Renoir. It is only by immersing oneself in the collected paintings in this incredible museum that one begins to understand Impressionist art.

The life stories of the engineers below are similarly meant to define what we mean by military engineering. The stories are grouped into three categories:

1. engineers who were in the armed forces,
2. engineers who worked in armaments production, research, or other activity directly funded by the military, and
3. engineers who believed that they were working on civilian projects but found that their work was used for military purposes.

The stories illustrate how military engineers are integral to the society in which they live and work and how they are indispensable to the leaders of their nations when these leaders want to wage war. In reading these stories, it might be useful to reflect on how these engineers interpreted their responsibility to the public. Did they understand what they were doing and then rationalize their actions, or did they ignore the implications and results of their work? If, at the conclusion of their professional lives, they had occasion to reflect on their life's work, would they have had any regrets or second thoughts?

2.1 Exemplars of Military Engineering

2.1.1 Engineers Who Were in the Military and Worked Directly for the Armed Forces

As noted in Chapter 1, armies need engineers. In the USA, the military academies emphasize engineering, and often the best and the brightest, on graduation, take commissions in the US Army Corps of Engineers or in similar organizations in the other armed services. This is military engineering in its historically most pure form.

Engineers working directly for the military are not free to choose the work they do. If they do not like their work, their only alternative is to resign their commissions, but even this is not an option at wartime. Moral conflicts can therefore be intense when the military engineer is asked to do something that he or she finds to be morally abhorrent. Military engineers often have to balance their commitment to their military service with their commitment to serve the public good.

In James A. Michener's book *Tales from the South Pacific* [8], a lieutenant on a small island toward the end of the Second World War is told to construct a landing strip. The construction requires cutting down a strand of stately Norfolk pines that are protected by an elderly islander. The lieutenant knows that the war is almost over and there really is no need for the airstrip, or for cutting down the strand of trees, but he has been told to do so. He has a moral conflict, as a commissioned military engineer, to follow orders on the one hand or to do what he perceives as being in the best interests of the public (and the trees!) on the other. Some of the engineers described below may have experienced similar conflicts, and some certainly did not. The common thread in all of these stories is that these engineers were engaged in military engineering.

Thaddeus Kosciuszko (1746–1817)

Fig. 2.1 Thaddeus Kosciuszko

Kosciuszko, born in 1746 in a small Lithuanian town, went to the military academy in Warsaw and graduated with an engineering degree. He fell in love with the daughter of the Marquis of Lithuania, but when his advances were not reciprocated, he left Poland and offered his assistance to the rebel army in the American colonies. He arrived with a letter of recommendation from Benjamin Franklin, addressed to General George Washington. Upon meeting Kosciuszko, Washington asked him what he could do. “Try me,” was the answer.

Kosciuszko received his commission and went to work as a military engineer. Horatio Gates, a general in the Revolutionary army, described Kosciuszko as an able engineer and one of the best and neatest draftsmen he had ever seen. Kosciuszko planned and constructed the encampment at Berets Heights, near Saratoga, which held out against the British attack, thus destroying the British strategy of cutting off New England from the rest of the colonies. Kosciuszko became the principal engineer in the construction of the fort at West Point, and the superior fortress assured the control of the Hudson River. He eventually became Washington’s adjutant and was promoted to brigadier-general at the end of the Revolutionary War.

With his mission accomplished, he returned to Poland where he lived several years in retirement, but then the Russians threatened to invade Poland and Kosciuszko was recruited to the defense of independent Poland. When the Russians overwhelmed the Polish army, Kosciuszko resigned his commission and went to live in France. The Russian occupation of Poland was, however, unacceptable and he became involved in a plot to drive out the Russians. He suddenly appeared in Krakow and issued a declaration of independence. He organized an army made up of mostly poor peasants with few weapons but they nevertheless routed the Russian army twice their size. Eventually, however, the Russians defeated the Poles, who retreated to Warsaw where they continued to defend the city against both the Russians and the Prussians. When the Austrians also attacked Poland, the

Polish army was defeated and the nation of Poland ceased to exist. Kosciuszko was wounded, captured, and imprisoned in St. Petersburg. After the death of Catherine the Great, the new tsar released him from prison and offered him his sword. Kosciuszko refused to accept it, saying, “I have no need of a sword; I have no country to defend.”

Kosciuszko visited the USA in 1797 and was received as a hero. Even though Congress gave him a grant of land and a pension, he decided to go back to Europe, where he spent the rest of his life in retirement, continuing to agitate for the independence of Poland.

Kosciuszko was not only a military engineer but a gallant soldier as well who seemed to always be fighting on the side of liberty and independence. He was buried in Wawel Cathedral (*Katedra Wawelska*) near Krakow, and the people constructed a huge mound nearby using earth brought from every great battlefield in Poland. A monument was erected to his memory at West Point by the US Military Academy cadet corps of 1828 [9].

Curtis LeMay (1906–1990)

Fig. 2.2 Curtis LeMay

Sometimes engineers are asked to use their knowledge and skills directly in line command, and often they are highly successful. One of the most successful engineers in combat was General Curtis LeMay.

Curtis LeMay studied civil engineering at Ohio State University and in 1930 became a second lieutenant in the US Army Air Corps. As WWII approached, he moved quickly up the chain of command, using his engineering education and skills to become a lieutenant colonel at the start of America’s involvement in Europe. He was transferred to the bombardment division and in 1944 to the Pacific theater as a major general in charge of the bomber command.

As the USA began to island hop toward Japan, air bases were established from which bombers could attack the Japanese mainland. LeMay received a new bomber, the B-29, designed for high-altitude bombing, and he was pressed by his superiors to achieve “results.” But the problem was that the bombers had to fly at altitudes of over 30,000 feet to be clear of antiaircraft fire, and thus their bombs were largely ineffective. LeMay discovered that Japanese antiaircraft guns were designed for defending against airplanes at high altitudes and could not be swiveled quickly. If an airplane came in at a low altitude, such as 3000 feet, the antiaircraft guns could not be turned quickly enough to fire effectively at the airplane. In addition, such low-level attacks allowed pilots to concentrate bombs accurately on small areas.

Most Japanese cities were susceptible to fire, and LeMay decided to drop incendiary bombs that would create huge firestorms in cities. He learned this technique from the British, who unwittingly had created such a firestorm in Hamburg, Germany, and then had intentionally firebombed Dresden where the entire city and its inhabitants were incinerated.

The new technique was applied to Japan with great success, and city after city was destroyed. On 9 May 1945 LeMay's bombers firebombed Tokyo, killing at least 100,000 people in one night. LeMay justified these attacks by arguing that the Japanese had decentralized their war manufacturing, and entire cities therefore became legitimate military targets. By the time the atomic bombs were dropped on Hiroshima and Nagasaki, there were no cities left to bomb on the Japanese islands since all significant population centers had been firebombed, with the loss of perhaps over 1,000,000 lives.

At the end of the war, LeMay observed that it was a good thing his side had won. Otherwise he believed that he would certainly have been tried as a war criminal.

Lucius Clay (1897–1978)

Fig. 2.3 Lucius Clay

Lucius D. Clay entered West Point in 1915, intending to become an army engineer. After graduation he taught civil and military engineering at West Point for 4 years. Before the Second World War he worked on several domestic construction projects and then served on the staff of General Douglas McArthur. In 1937 he became responsible for enlarging and improving military airports in the USA.

During the Second World War Clay became the head of supply for all of the armed services. His most remarkable accomplishment occurred during the Normandy invasion. He was sent by General Dwight Eisenhower to reopen the Port of Cherbourg so that critical supplies could be shipped to the troops. The Germans had destroyed the port facilities and the port was in chaos. Nevertheless, he was able to stabilize the situation and within one day the port was again operating with speed and efficiency. His contribution to the success of the D-Day landings was significant and he was awarded the Bronze Star for his efforts. His most notable achievement, however, occurred after the war. He became the governor general of Germany and set about helping Germany get back on its feet. He worried about the division of Germany into four zones, and mostly about the fate of those Berliners who were trapped in the Russian zone. When Soviet harassment of war-torn Berlin began in early 1948, Clay acted on his own initiative and ordered an airlift to supply Berlin. Only after the operation was already in full swing did he convince his superiors that Berlin should be supplied by air. His optimism and hard work helped the USA to carry out the dangerous mission of flying into the Berlin airport with vital supplies for the beleaguered city.

General Clay retired in 1949, within days after the Soviet blockade of Berlin had been lifted. To Berliners he was a hero for having broken the Russian blockage, and a wide boulevard in the middle of the city is now named Clayallee in his honor. General Clay was buried in the cadet cemetery at West Point [10].

George Goethals (1858–1928)

Fig. 2.4 George Goethals

George Washington Goethals studied engineering at the City College of New York but graduated from West Point, second in his class. On graduation he returned to West Point as an instructor in engineering. During the Spanish-American War he served as chief of engineers and was later placed in charge of the Muscle Shoals canal construction on the Tennessee River. In 1907 received the assignment of a lifetime: President Teddy Roosevelt told him to build the Panama Canal.

As chief engineer in charge of canal construction he faced many obstacles, only some of which had to do with construction. Heat and tropical diseases took their toll on workers and many doubted if the canal could ever be finished. But by 1915 the job was done and Goethals stayed on to serve as the Governor of the Panama Canal. In the USA he was universally praised for this engineering feat and was even awarded an honorary doctorate from the University of Pennsylvania.

When America entered the First World War he was appointed quartermaster general and put in charge of all materiel used by the U.S. army. After the war he resigned his commission and went into private practice, which included serving as a consulting engineer to the Port Authority of New York.

The Panama Canal became immensely important to the USA during the Second World War when troops and materiel had to be transferred between the two oceans. Fighting a war on two oceans would have been impossible if troops and supplies could not have been easily transported by sea through the canal. What was originally a commercial venture became a most crucial advantage to the US military.

Yahya Ayyash (1966–1996)

Fig. 2.5 Yahya Ayyash

Yahya Ayyash was not in any official army, did not wear a uniform, and did not get paid for his work, but he was a military engineer.

Yahya Ayyash did not build anything useful, but he nevertheless became known as "The Engineer" and was adored by thousands to whom he was a hero. What Ayyash did was build bombs. He worked with the Palestinian group Hamas, which during 1994 and 1995 organized suicide attacks on Israel. Ayyash is alleged to have participated in 11 suicide attacks, killing from 50 to 75 people, with hundreds more wounded.

Ayyash was from all accounts an exemplary person with a talent for engineering. He was born in 1966 in the Gaza Strip, a barren, impoverished strip of sand created after Israeli independence to hold displaced Palestinians. After high school, Ayyash studied chemical engineering at Bir Zeit University. He apparently was a devout scholar of the Koran in addition to his work in engineering. After graduation he married and had two sons.

His experiences in Gaza during his childhood and his dedication to the Islamic cause led to his becoming active in Hamas, which in the mid-1990s was a terrorist organization dedicated to the destruction of Israel. In April 1994 he built his first bomb, which was used in an attack on a bus station in northern Israel, killing nine people. As his effectiveness as a builder of bombs increased, he became a priority target for Israeli intelligence, which finally caught up with him in 1996. He was given a new cell phone by an uncle of a friend, and as soon as he tried to use it, it exploded, killing him instantly. His funeral was attended by 100,000 people and he was declared a martyr by Hamas [11].

2.1.2 Engineers Who Worked Directly for the Military

Many engineers work indirectly for the military, doing research on armaments or consulting with the military. Their income comes from the military establishment, but they have the option to not work on projects they consider immoral. Their military engineering work differs from those who are officers in the services because they are not required to follow orders. Their cost for doing the right thing is far less than for those in active military service, and for this reason they have a greater responsibility to society.

Vauban (1633–1707)

Fig. 2.6 Vauban

During the 17th century, the geographical position of France required it to fortify its borders, and a series of forts was constructed in its periphery. The work of building these forts was entrusted to Sebastian le Prestre de Vauban, who had distinguished himself both as a soldier and engineer. Vauban used the idea of the star-shaped fort originally developed by the Italians (Chapter 1) and refined these so that they were thought to be almost impregnable.

Although Vauban built the forts, this was a case of a fox building a chicken coop, because Vauban soon figured out ways of attacking the very same impregnable forts he had designed. His plan for attacking the forts involved digging ditches in parallel arcs, constantly moving closer to the fort. The diggers piled the dirt between them and the fort and thus were protected by the very earth they were digging out. Artillery was then brought forward into the trenches, again protected by the earth embankment. Eventually the attackers were able to get so close that cannons could obliterate a section of the wall and the infantry could storm the relatively low and now demolished walls. This technique was used by General George Washington in the Battle of Yorktown and ensured the survival of the new United States of America. Vauban's trench technique remained a system of siege warfare through the First World War.

While Vauban is remembered as a great engineer, his single most important contribution was a short pamphlet he wrote that foreshadowed the French Revolution. He argued that the deplorable condition of the peasants was unacceptable and that a tax must be levied on all persons, including the nobility, if the defense of France was to be assured. King Louis XVIII immediately suppressed the document and had Vauban banished from court. After Vauban died, his reputation was restored by Napoleon, who had Vauban's heart reburied in the church at Invalides.

Wernher von Braun (1912–1977)

Fig. 2.7 Wernher von Braun

A characteristic of military engineering, both in days past and today, is that the engineer often is not concerned on whose side he or she works. A classic case is Wernher von Braun.

While von Braun was a student at the Institute of Technology in Berlin he became interested in rockets and worked in his spare time with Germany's most imaginative scientists. He graduated in 1932 with a degree in aeronautical engineering and looked forward to a career in aviation, including getting his pilot's license. He soon became convinced that if rocketry were to be used to explore space, more advanced technology would be needed, and he enrolled at the University of Berlin to work toward a PhD in physics. His dissertation was on the use of liquid fuels in rockets, then a new concept. After graduating in 1934 he was hired by the German Army Ordnance Corps where he conducted experiments using liquid fuel rockets. When war came, von Braun continued his work as a civilian employee for the German army. He designed the dreaded V-2 rocket that terrorized London during the last months of the war. After Germany's capitulation, von Braun and 120 of his colleagues surrendered to US forces in order to avoid capture by the Russians.

The German scientists and engineers were brought to the United States and began their work for the US army, designing and building the Redstone, Jupiter, Juno, and Saturn rockets. In 1957, one of von Braun's rockets put Explorer 1, the first American satellite, into orbit, after an embarrassing misfire and explosion by the Navy's Vanguard rocket. Von Braun's rockets were used for the Mercury manned missions into space, and his Saturn rockets lifted the Apollo teams into orbit. When President Kennedy declared our intention to put a man on the moon, von Braun was asked to build the rocket that would get them there. His Saturn V was highly successful and is still used, decades later, for the space shuttles.

The most important use of von Braun's rockets was to serve as vehicles for carrying nuclear warheads, and thousands were built and stored in silos all over the USA. The rockets were thought to be a deterrent against attack by the Soviet Union and remained active throughout the Cold War, from about 1950 to about 1990. The USA continues to have an unspecified number of these rockets in underground silos, armed with nuclear warheads, to be used in case of attack.

The ease with which Wernher von Braun became the leading rocket expert for the USA after serving in the same capacity for Germany during the Second World

War illustrates how some engineers concentrate on their jobs without being concerned with what their efforts can produce. During the 1960s Tom Lehrer, a satirist and also a mathematics instructor at Harvard, recorded some wonderful songs that spoke to the craziness of the age. One of his songs was about Wernher von Braun; it goes like this:

Gather 'round as I sing you of Wernher von Braun,
A man whose allegiance
is ruled by expedience.
Call him a Nazi and he won't even frown.
"Ha, Nazi, Schmazi," says Wernher von Braun.
Don't say that he's hypocritical,
Say rather that he's apolitical.
"Vonce zee rockets are up, who cares vere zey come down?
Zat's not my department," says Wernher von Braun.

Some have harsh words for this man of renown,
But some think our attitude
Should be one of gratitude,
Like the widows and cripples of old London town,
Who owe their large pensions to Wernher von Braun.

You too can be such a hero
If you can count backward to zero.

"In English or German, I know how to count down,
Und I'm learning Chinese!" says Wernher von Braun. [12] [T. Lehrer, pers. comm.]

John Napier (1550–1617)

Fig. 2.8 John Napier

Learned men in the 1500s, such as John Napier, might have been engineers if that designation had existed. Their work was certainly in military engineering. While better known as a mathematician and the inventor of logarithms, John Napier also proposed military inventions, including burning mirrors that set enemy ships on fire, artillery that destroyed everything within a radius of 4 miles, bulletproof clothing, a crude version of a tank, and a submarinelike device. We have no evidence that he actually developed any of these, and he took great pains to conceal the details of his inventions. On his deathbed in 1617 he worried about the development of more advanced armaments. He was a reluctant military engineer.

Kurt Prüfer (1891–1952)

Fig. 2.9 Kurt Prüfer

Sometimes the tasks given to engineers during wartime are so horrific that it is difficult to understand why the engineers simply do not refuse to do the work. One such case, and perhaps one of the most horrible cases, is Kurt Prüfer.

Kurt Prüfer was a German civil engineer who, after returning from the First World War, went to work for Topf and Sons, a venerable company founded in 1878 that made customized incinerators and malting equipment. In the 1920s, under Prüfer's direction, the firm pioneered in the design and construction of clean crematoria, furnaces that complied with strict regulations on preserving the dignity of the body. No flame was to come into contact with the coffin and the process had to be smoke and odor free. In 1928 Prüfer was promoted to head the crematorium-construction division. The Great Depression strained the viability of the business, but by 1934 conditions improved and Prüfer retained his job. He further strengthened his position with the firm when he joined the Nazi party.

The firm was located close to what later became the Buchenwald concentration camp and had attained such a good reputation for quality work in the construction of crematoria that in 1939 the Nazi SS approached Prüfer with an order for a crematorium. An epidemic at Buchenwald had allegedly killed hundreds of prisoners, and an efficient furnace was needed to cremate the remains. Prüfer designed a crematorium resembling an incinerator used for animal carcasses, knowing that the dead were not to be burned individually or in coffins.

The furnaces were quite successful, and orders came rolling in from the other concentration camps. It ought to have been quite clear to Prüfer that the furnaces at these camps were not being used for burning people killed by epidemics. Prüfer even visited Auschwitz several times and saw that his ovens were being used for "special operations." Rather than recoiling in disgust, Prüfer accepted the engineering challenge to develop increasingly efficient means of extermination.

After the war he was captured by Americans but then inexplicably released. The Russians then arrested him and tried him for war crimes, sentencing him to 25 years in prison. The transcripts recorded during his interrogation show he felt no remorse for what he had done. In fact, after the Nazis abandoned Auschwitz in January 1945, Prüfer suggested to the Nazis that they could reassemble parts of the furnaces in Mauthausen concentration camp in Austria so that they could continue with their program of extermination.

During the interview with the Russian interrogators, this exchange occurred:

> Q. What motivated you to continue with the building of the other crematoriums as senior engineer with Topf and Sons?
> R. I had my contract with the Topf firm and I was aware of the fact that my work was of great importance for the national socialist state. I knew that if I refused to continue with this work, I would be liquidated by the Gestapo [12].

His response is important in understanding his commitment to society. The first reaction was that he was just doing his job and thus was absolved of all blame. Perhaps he then recognized that this was not a valid excuse, and rationalized his actions by saying that he had no choice but to go along with the Nazis. He was, however, a member of the Nazi party and enthusiastically participated in the exterminations, so this is hardly a legitimate excuse.

After the war the firm of Topf and Sons was nationalized by the East Germans and remained in business until 1996.

Gerald Bull (1928–1990)

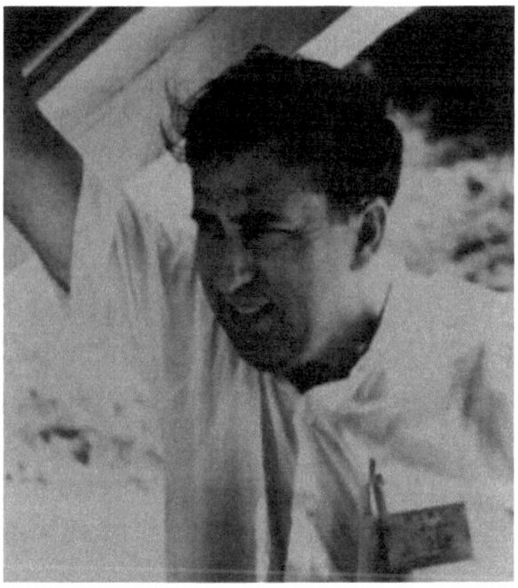

Fig. 2.10 Gerald Bull

One of Gerald Bull's biographers said of him: "Oddly enough, Bull was not a militarist. He never saw military service or even owned a handgun. He was said to be generous and thoughtful. He just found something that he really loved doing" [13]. What he loved doing was building really big guns, guns that could achieve muzzle velocities great enough to place satellites into orbit, or send missiles to targets thousands of miles away.

Bull, born in 1928, was a Canadian who did not have a happy childhood, but who excelled in school, obtaining his engineering degree and then, at age 23, a PhD in engineering from the University of Toronto. He went to work for the Canadian Armament and Research Development Establishment (CADRE), which, before the Second World War, was involved in bringing German scientists and engineers to Canada in order to deprive the Nazis of their talents. Some of their research involved supersonic aircraft, and they needed to construct a large wind tunnel to test the missiles. This was very expensive, and Bull hit on the idea of testing supersonic flight by shooting the missile out of a large gun. The idea worked brilliantly, and this work encouraged Bull to continue work on large guns. But Bull was apparently not a very social person and alienated the research establishment, which finally eliminated his funding.

Undeterred, he set up a private company to continue the research and looked around for anyone willing to pay money for his knowledge. South Africa, which was fighting Angolan rebels at the time, hired him to build a gun that could shoot long distances. Bull built and sold them hundreds of 155 howitzers that could shoot shells 50% farther than any other artillery in use at the time. The sale, however, got Bull in trouble for illegal arms dealing, and he served 6 months in an American jail. When he got out of jail he was broke and had few options, so he began to sell his expertise to anyone willing to buy. Iraq at the time was at war with Iran and hired Bull to build the howitzers it used with devastating effect in the Desert War. Part of the deal with Iraq was that Bull could also continue his research on large guns capable of sending satellites into space (or nuclear warheads into Israel). Iraq was interested in the possibility of launching missiles using the big guns, but before this project could be completed, Bull was assassinated by Israeli agents, and the supergun project was abandoned [14].

Gerald Bull became so involved in his work that he became estranged from Canada, went to prison in the USA, and eventually was assassinated by Israel. His vision was of an entirely new way to get into space – the use of giant guns to send the payloads into orbit. To achieve this objective he worked at various times for some of the least savory regimes on earth: apartheid South Africa, Communist China, and, ultimately, Iraq. His work might have had a devastatingly destabilizing effect in the Middle East and could have resulted in the deaths of millions of people [15]. Did Bull ever think of this when he was happily building his big guns?

Galileo Galilei (1564–1642)

Fig. 2.11 Galileo Galilei

Galileo was often strapped for cash and went into the private tutoring business, teaching all comers everything he knew about military architecture, fortifications, and surveying. In 1597 a lot of foreigners lived in Padua, and his tutoring business was good.

He was an inveterate inventor and developed many gadgets that were useful for the military, such as a "geometric compass," which made possible the rapid calculations of geometric shapes, for example estimating the midpoint of a line. The device, which could be thought of as a primitive slide rule, is simple in construction but requires instruction in its use. Galileo set up a company that sold these instruments and hired craftsmen to construct them. He held seminars explaining their use, especially in warfare. During this time many of the city-states in Italy were at war with each other, and the pope was fighting with most every one of them, so Galileo's business prospered. Much has been written about Galileo's scientific discoveries and contributions to basic knowledge but the source of his income came from military engineering. He did not seem to have much moral concern about doing military engineering as long as it paid well. The only instance when he seemed to care about killing others was when he "invented" the submarine. He decided not to publish his invention because he feared that this would cause too much harm.

Galileo is well known for his arguments with the Church during which he raised all kinds of moral concerns, but except for not divulging the plans for the submarine, he did not have any trouble selling his expertise to those who would use the information for warfare. He was just a military engineer doing his clients' bidding [16, 17].

Michael Cantrell (1955–)

Fig. 2.12 Michael Cantrell

A more recent example of an engineer who, while working on military projects, did well for himself is Michael Cantrell. Cantrell worked for the Army Space and Missile Defense Command in Huntsville, AL, which was the central organization on the so-called "Star Wars" project initiated by President Ronald Reagan and continued under subsequent administrations. Even though there is every reason to believe that the entire concept of Star Wars is faulty, we have thus far spent over $110 billion, with no end in sight [18].

At the inception of Star Wars a great deal of money flowed into Huntsville, but as the idiocy of the project began to permeate military planning, the rate of money flow slowed down. Middle-level managers such as Michael Cantrell, in order to be able to continue their programs and to pay for their research and development facilities, had to compete with other defense agencies for public money. Cantrell emerged as a master at getting defense contracts and was able to find funding for projects that the Department of Defense did not even want to fund.

Though it was strictly against the rules for federal employees to lobby for work, Cantrell started to find ways of getting money for Huntsville. He set up an office at the US Airways lounge at the Washington National Airport and hid from his bosses, all the while arguing with lawmakers and congressional staff for money for his projects. He was able to get contractors to send contributions to the election funds of congressmen and then called their offices to collect on the favors. He paid lobbyists to make sure the projects were funded, and he counted on the lawmakers to protect him and his activities. The oversight from the Pentagon was so lax that Cantrell was able to bamboozle the Department of Defense into paying for phantom projects. They would, for example, send a section of a rocket back and forth between two contractors and charge the government each time for the construction costs.

Getting money to Huntsville was so easy that Cantrell and another engineering colleague decided to tap into the funds. They contrived to bill the government for work not done and to pocket the funds, and they were highly successful. Any concerns expressed by those who recognized the scam were rapidly quelled by powerful congressmen who received money from Cantrell or from contractors. With his new-found riches, Cantrell built a mansion in an exclusive neighborhood in Huntsville and lived the high life. Eventually, however, the investigators caught up with him. He was convicted of conspiracy and bribery and is now in jail [19].

Simon Stevin (1548/49–1620)

Fig. 2.13 Simon Stevin

Little is known of Simon Stevin's childhood or of his education. He was brought up in what is today the Netherlands in the Calvinist tradition. He attended the university at Leiden, matriculating at age 35.

In the middle 1500s the states in the low countries were rebelling against the Spanish, who were trying to control the region. While at the university, Stevin met and befriended Maurits, the second son of William of Orange. When William was assassinated, Maurits became the head of the army and the republic of the Netherlands, and Stevin became his advisor, helping Maurits beat back the Spanish in several key battles. Maurits understood the importance of military strategy, tactics, and engineering and asked Stevin to set up an engineering school within the University of Leiden. During these years, Stevin invented a way of flooding the lowlands in the path of an invading army by opening selected sluices in dikes, and he advised on the building of windmills, locks, and ports. He helped Prince Maurits build fortifications for the war against Spain and wrote detailed descriptions of military innovations that were adopted by the army.

In addition to being the most prominent military engineer of the emerging Dutch nation, he also authored 11 books, making significant contributions to trigonometry, mechanics, architecture, music theory, geography, fortification, and navigation. In 1585 he published a booklet that introduced decimal fractions, a development that greatly benefited trade and manufacturing. He also wrote important works on mechanics, mainly dealing with hydrostatics, recognizing for the first time that the pressure exerted by a liquid upon a surface depends on the height of the liquid and the area of the surface. He also beat Galileo to the punch (by 3 years) by conducting experiments involving dropping lead balls of different weights and finding that, when dropped from the Delft church steeple, they hit the ground at the same time. He also contributed to the science of music, for the first time explaining the correct theory of the division of the octave into 12 equal intervals [20].

The most significant contribution Stevin made was his work in support of his country and his people. The independent state of the Netherlands would not have been possible without the work of Simon Stevin, who is now venerated as a national hero.

Abdul Khan (1936–)

Fig. 2.14 Abdul Khan

Abdul Quadeer Khan was born in Bhopal, India in 1936 to a Muslim family. At that time the population of Bhopal was made up of both Muslims and Hindus who lived an uneasy but peaceful coexistence. When India was partitioned, forming Pakistan, a huge migration occurred with Muslims moving north to Pakistan and Hindus moving south to India. Abdul Khan, one of seven children, moved with his family. It was a cruel trip. The Khan family was harassed, beaten, and robbed, and Abdul Khan ended up walking barefoot to Pakistan. The experience forever impressed on him a distrust and hatred of Hindus and India.

Khan studied at the university in Pakistan and then at the Catholic University in Belgium. After graduating with a PhD in metallurgical engineering, he was employed in the Netherlands by a German company that manufactured high-speed centrifuges used in the production of weapons-grade uranium. India had developed and demonstrated its nuclear bombs and was threatening the newly ceded Pakistan. Taking advantage of an insurrection in eastern Pakistan, the Indian army had already soundly defeated the Pakistani forces, resulting in the formation of a new country, Bangladesh.

Recognizing the perilous situation, the Pakistani dictators decided that they also needed nuclear weapons. Part of the plan was to have Khan send classified information from his job in the Netherlands on the processes used to enrich uranium. Khan was highly successful in his spying, but eventually his activities were discovered and he had to flee to Pakistan, where he was given the job of organizing a facility for producing nuclear bombs and the missiles to carry them. Using his engineering skills and an unlimited budget he soon succeeded in producing a nuclear weapon and became a national hero, living a life of privilege and amassing honors and adulations.

It is unclear whether he or the government of Pakistan decided that there was money to be made in sharing his knowledge for producing nuclear weapons. At any rate, Khan began to sell this knowledge on the manufacture of nuclear weapons to rogue nations such as Lybia, Syria, and North Korea. Apparently even al-Qaeda tried to get a nuclear weapon from Khan. In 2001 the Pakistani government arrested three nuclear scientists who had close ties to Khan for trying to get nuclear weapons to the Taliban in Afghanistan. Khan's activities were soon revealed, much to the embarrassment of the government of Pakistan, which decided to make Khan the scapegoat, arresting him and keeping him in detention and incommunicado [21]. He was given limited freedom in 2009.

Khan's actions demonstrate how easy it is for a talented engineer to act immorally and with little regard for the public – in this case providing nuclear weapons to countries that may have every intention of using them to terrorize the world.

Fritz Haber (1868–1934)

Fig. 2.15 Fritz Haber

Fritz Haber was born into a well-to-do German-Jewish family. He studied science and engineering at several German universities and upon graduation took an appointment at the Department of Chemical and Fuel Technology at the Polytechnic in Karlsruhe, Germany. In 1911 he became director of the Institute for Physical Chemistry and Electrochemistry at the new Kaiser Wilhelm Gesellschaft in Berlin, where academic scientists, government, and industry cooperated to promote original research.

During the First World War the British blockade prevented Germany from importing ammonia nitrogen, which was crucial in the manufacture of gunpowder. This could have spelled the end of the war, but Fritz Haber figured out how to fix nitrogen from the air, and, using high pressure and a catalyst, he was able to produce ammonia. The process was soon scaled up by the scientists working at the huge chemical firm BASF. One of their scientists, Carl Bosch, was able to apply it to the manufacture of ammonia nitrogen and the process became known as the Haber–Bosch process. Germany was able to produce unlimited gunpowder, and this development was responsible for the loss of countless lives in the prolonging of the war.

Haber continued to use chemistry to help his country win the war in whatever way he could, and one of his contributions was the development of a new weapon, poison gas. On 22 April 1915, near Ypes, Belgium, the Germans opened 6,000 cylinders of chlorine under the supervision of Fritz Haber himself and watched as the cloud of yellowish gas soon reached the French positions and suffocated the soldiers who had not fled in time. Because of his work with poison gas, Haber is often referred to as the "father of modern chemical warfare." Although perhaps 10,000 troops died from poison gas that day, the weapon did not prove to be decisive since the Allies soon learned to cope with it and retaliated with poison gas attacks of their own. The success of the poison gas weapon was not without personal cost, however. Haber's wife, Clara, who was also a chemist, became distraught when she found out about Haber's involvement in the use of poison gas and committed suicide.

In 1918 Haber was awarded the Nobel Prize for his work on the synthesis of ammonia. Given his involvement in the war, the prize was roundly criticized. When the Nazis came to power, Haber's situation in Germany became untenable because of his Jewish ancestry, and in 1933 he decided to leave Germany for Switzerland. He became an outcast from a country he loved and which he had worked so hard to support. His life has become a popular subject of many plays and books because of the intense moral conflicts he must have wrestled with. Or did he?

Henry Kaiser (1882–1967)

Fig. 2.16 Henry Kaiser

Henry J. Kaiser did not have an engineering education, but he had an exemplary engineering career. He got things done. Going to work when he was 13 years old, he eventually founded over 100 companies including Kaiser Aluminum, Kaiser Steel, and Kaiser Cement and Gypsum, and he created the first health maintenance organization (HMO) for his shipyard workers. After the Second World War he was one of the first to mass-produce cars for the public. The Kaiser and the Henry J (one of the first compacts) were initially wildly popular, but eventually the larger companies caught up with Kaiser and he was forced him to stop producing cars.

Kaiser's first career was in construction, beginning with a gravel company and ending with the formation of a consortium of companies that built the Hoover Dam, the Bonneville Dam, and the Grand Coulee Dam, as well as many other large projects.

During the Second World War, German submarines in the Atlantic were extracting a dreadful toll on American shipping, with more ships being lost than could be constructed. Kaiser's solution was to speed up the construction of ships. He introduced assembly-line production to his seven shipyards, and by the end of the war his yards had produced 1,490 ships. These so-called "liberty ships" were instrumental in carrying military cargo and troops overseas.

After the war, it became clear that the USA could not have become the military power it did in Europe without the ships Henry Kaiser had built. The overwhelming superiority of the American military in Europe was responsible for the lingering effect of American social influence even today, including the use of American English as the new international language of commerce and technology.

Jerry Baber (1937–)

Fig. 2.17 Jerry Baber with his robot helicopter. The shotguns are slung on the side

The rifle has been the infantry's weapon of choice for several hundred years, but its days may be numbered. A shotgun, spraying small pellets at very high velocities, is much more lethal at limited distances such as would occur in urban warfare. But the shotgun also has disadvantages. Its range is limited, and shotguns must be manually loaded, with shells placed one at a time into the barrel. Jerry Baber, an electrical engineering graduate of Virginia Tech, working out of his garage in Piney Flats, TN, figured out how to eliminate this problem and has manufactured a rapid-fire, fully automatic shotgun. Its most innovative feature is the absence of recoil, increasing its accuracy and preventing bruised shoulders. With this gun the bolt never bottoms out or slams into a fixed object at the end of its rearward travel, but stops gradually against the long recoil spring.

When Rototex, a manufacturer of small robots, wanted to install firepower on their battlefield robots, they called Jerry Baber. The result of the collaboration is a lethal and scary robotic tank, about the size of a medium dog that can move into a firefight and blast the enemy at 300 rounds a minute, more firepower than that of a platoon of soldiers. The weapon on the tank is made of stainless steel so it cannot rust and never needs lubricating, and the fact that it does not recoil makes it perfect for use in the robotic tank. Another one of Baber's innovations is a small helicopter with shotguns slung on the side. This unit can carry the robots into battle [22, 23].

One of Baber's most ardent supporters and customers is Blackwater, the North Carolina security firm that had the security contract in Iraq and is now fighting several lawsuits stemming from their insensitive treatment and alleged murder of Iraqi civilians.

Baber believes that the deploying of the robots will revolutionize warfare. Others, including the armed services, are not so sure. Robots in the battlefield have the advantage that no human being is being put at risk, but on the other hand, robots can be compromised and turned on the very people who initially deployed them. Having no brains and no morals, robots kill indiscriminantly, and thus their use in a combat situation can be problematic. Nevertheless, Jerry Baber believes that his gun will find wide use in modern warfare, whether carried by a robot or a living soldier [24].

Willy Messerschmitt (1898–1978)

Fig. 2.18 Willy Messerschmitt

The Battle of Britain defined the course of the Second World War. Once the British fighter squadrons in their Spitfires and Hurricanes showed that they could hold their own against the German Luftwaffe, it was clear who would win the war. The Royal Air Force denied air superiority to the Germans and forced Hitler to abandon the invasion of the British Isles. Victory in the Battle of Britain was not a foregone conclusion, however, due mostly to one airplane, the ME (for Messerschmitt) 109.

Willy Messerschmitt loved aviation from the time he was a little boy and went to a military flying school during the First World War. He then started to build gliders and small airplanes that were revolutionary in design but tended to crash easily. Then, working for the Bavarian Aircraft Works, he designed and produced a small transport plane that was cheap to build but also had stability problems. Nevertheless, when the Nazis came to power in 1933, Bavarian Aircraft Works was awarded several large contracts to build airplanes that could be used for military purposes, secretly contravening the Treaty of Versailles. When the Luftwaffe announced a contest for the design of a fighter plane, Messerschmitt and his colleagues designed what was to become the ME 109. It was a revolutionary aircraft, built with a metal frame, an enclosed cockpit, and a thin single wing and using a powerful water-cooled engine. The Luftwaffe agreed that the 109 was the best airplane that had ever been built and ordered large numbers of them. It became the very weapon the Germans needed to gain air superiority in Europe during 1939–1940. Although other airplanes such as the Focke–Wulf (FW) 190 was supposed to eventually replace the ME 109, these were never produced in sufficient numbers and the ME 109 remained the first-line fighter for the Luftwaffe. It was an effective match for the British Spitfire with a higher ceiling and better armaments.

Fig. 2.19 The ME 109

Messerschmitt did not rest on his laurels and went on to design other aircraft such as the ME 110, a twin-engine, two-seat fighter that was an effective weapon against Allied bombers. His ME 210, however, was a disaster. The airplane was supposed to be a new twin-engine bomber, but Messerschmitt insisted on reducing the weight of the airplane in order to increase its speed, but in so doing, its ruggedness was reduced. The biggest problem was that the undercarriage tended to collapse on landing, causing catastrophic crashes. Relying on Messerschmitt to produce useful airplanes, the Luftwaffe had ordered over 1000 of the ME 210s "off the drawing board" without prior shakedowns and flight tests, and with the failure of the plane, they did not have a replacement for the Heinkels and other small bombers being shot down by the British.

Toward the end of the war Messerschmitt started to experiment with jet engines and produced a jet fighter, the ME 163, that was vastly superior to the piston-powered airplanes, but by then the production of aircraft in Germany had slowed to the point where the jet had little overall effect on the outcome of the war. Messerschmitt also had on his drawing board a large bomber that would have had the capability of crossing the Atlantic Ocean and bombing the USA, but this aircraft was never built.

After the war, the Bavarian Aircraft Works was merged into a large conglomerate and went into the business of building cars as well as working in the design of transport planes for the European consortium. Messerschmitt did not design another airplane, although he helped the aircraft industries in other countries with the development of their own air forces.

Was Willy Messerschmitt an engineering design genius, or was he simply a good organizer who got others to do the design work? Certainly he left behind a legacy of unsuccessful airplanes, and perhaps no engineer has ever been responsible for more crashes during shakedowns and flight tests. What is true is that he loved aviation, and the war and the Nazis allowed him the opportunity to apply his engineering skills to the manufacture of airplanes [25].

2.1.3 *Engineers Who Worked in the Civilian Sector but Whose Work Became Useful to the Military*

Vannevar Bush (1890–1974)

Fig. 2.20 Vannevar Bush

Vannevar (van-NEE-var) Bush has been called the "engineer of the American century," and rightfully so. The present military/government/industry/university research establishment in America is largely his doing.

Born in Chelsea, MA, to Unitarian parents, he did well in school and developed a self-confidence he was to carry throughout his life. He went to Tufts University to study engineering and earned a master's degree in the same time it usually takes to finish the bachelor's degree. After graduation he went to work for General Electric but, after being laid off, took a teaching position at Clark University. The next stop was MIT, where he earned a PhD in 1 year. He then returned to Tufts as an assistant professor. By this time the USA was about to enter the First World War, and Bush decided to aid the war effort. He invented a device that used magnetic fields to detect submarines, and although this instrument was successful in field trials, the war ended before it could be effectively deployed on ships.

In 1919 Bush accepted a position in the electrical engineering department at MIT and started work on what we now call analog computers. By 1931 he had invented an analyzer that could solve differential equations. The idea of data storage and rapid retrieval fascinated him, and he developed a microfilm device that could review 1,000 fingerprints a minute, but the FBI was not interested in his research. In 1937 he became the president of the Carnegie Institution which conducted contract research, and the prestige of this position gave him access to governmental research laboratories.

The USA was woefully unprepared when it entered the Second World War. Military research was done in governmental labs and their output was dismal. Many scientists and engineers felt that an organization was needed that could do independent work, but with government research funding. In a meeting with President Roosevelt, Bush proposed a new organization for conducting research.

The president agreed, and the National Defense Research Committee was formed, with Bush as its head. The initial purpose of the organization was to coordinate military research, which led to the creation of the Office of Scientific Research and Development, which ultimately conducted most of the military research during the war. Some of the innovations from these laboratories during the Second World War were improvements in radar, the development of the proximity fuse for antiaircraft guns, and improvements in submarine tactics. Most of the work done was, of course, secret, including work on the Manhattan Project.

After the war Bush pushed for what became the National Science Foundation. Bush's idea was that all military research funding would be run through this organization, but this did not occur and today the NSF remains a minor player in military research. But Bush's most important victory was in having much of this funding go to universities, which, prior to the Second World War, had not been players in military research [26].

Bush's own research in memory retrieval led to something he called "memex," a storage and retrieval device using microfilm that could extend the powers of human memory and association. His device projected images on a screen, allowing the researcher to make associations and extensions of ideas, establishing links between documents just as the human mind forms memories through associations. The notion was the forerunner of what today we call hypertext, and thus Bush is often credited as one of the founders of the Internet.

Bush's effort without doubt helped to win the Second World War, but more than that, he changed the way military (and related) research was conducted. He was responsible for creating the system that defines the way military research is funded today [27].

Gaspard Monge (1746–1818)

Fig. 2.21 Gaspard Monge

As discussed in Chapter 1, during the 17th and 18th centuries the best military engineers were French. In addition to the development of shrapnel shells, pontoon bridges, gun carriages, and the semaphore signaling system, the French invented descriptive geometry, which was essential in the description of three-dimensional objects on two-dimensional paper, leading to what we today call technical drawing. The invention of orthographic projection is credited to Gaspard Monge, but the French military recognized the importance of this invention in the construction of military hardware and fortifications and prevented the invention from being published. Monge had to wait 25 years before his ideas became public.

Gaspard Monge was born in 1746 into a poor household. He was educated in schools that emphasized science and mathematics, skills that were useful to the military. Even though Monge was a brilliant student, he was not of noble birth and therefore could not be admitted to officer training. Instead, he was sent to a technical school to learn surveying and drawing. While he was a student, he was asked to plan a fortress based on surveying data, and while solving this problem, Monge invented the concept of orthographic projection. His finished drawings were at first rejected because they were done too fast, but Monge was able to show how his new technique was superior to the old slow way. He was so persuasive that he was soon appointed professor at the very school where he had been studying, but only on the condition that the principles of orthographic projection be kept a military secret and shared only with officers above a certain rank. In 1780 he was appointed to a chaired professorship in mathematics in Paris where he published an influential paper on curvature, a problem that had first been considered by Euler, the Prussian mathematician.

Monge was an early supporter of the French revolution, but during The Terror he was denounced and escaped the guillotine only by a hasty flight from Paris. He eventually joined Napoleon's armies in Egypt, where he watched from land as the French fleet was destroyed by the British. He escaped back to France and was made professor at L'Ecole Polytechnique, where he gave lectures on descriptive geometry. While he thought of himself as a mathematician, his development of the orthographic projection technique contributed significantly to military engineering where three-dimensional objects such as fortifications and weapons needed to be expressed in two dimensions on paper so that they could be reproduced in quantity.

Garrett Morgan (1877–1963)

Fig. 2.22 Garrett Morgan

Garrett A. Morgan was born in Kentucky to former slaves and spent his childhood on the family farm. When he was 14 years old he went to Cincinnati in search of employment. The only work he could find was menial handyman work, but he made enough to hire his own tutor, thus continuing his education.

In 1907 Morgan opened his own sewing machine and shoe repair shop and later expanded his business to include a tailor shop that made coats and suits, all with equipment that he had designed and constructed. In 1920, Morgan moved into the newspaper business, establishing *The Cleveland Call* and becoming a prosperous and widely respected businessman.

Morgan continued to invent things, including a safety hood and smoke protector. His "safety hood" was patented as a "breathing device" and consisted of a hood worn over the head of a person onto which was attached tubes that reached near the ground, bringing in clean air. The bottom of the tube was lined with a sponge-type material that helped to filter the air. He sold his invention around the country, although in many instances he was forced to allow a white partner to take credit for the invention. He became famous when he and his brother used the safety hood to rescue 32 men trapped during an explosion in an underground tunnel 250 feet beneath Lake Erie. The tunnel had filled with poison gas, and several men had died trying to reach the workers. Morgan, his brother, and a team of volunteers donned the new "gas masks" and went to the rescue.

Although other kinds of gas masks had been invented, the heroic efforts in saving the men trapped in the tunnel explosion was great publicity and the Morgan safety hood became widely used by fire and rescue workers. A refined model of the mask won a gold medal at the International Exposition of Sanitation and Safety and another gold medal from the International Association of Fire Chiefs.

When Morgan invented the "safety hood" he could never have imagined that the most important use of his mask, now much improved, would be as a lifesaving device during the First World War. The British version of the respirator was distributed a month before the Battle of the Somme and its use greatly reduced casualties from German poison gas attacks.

Fig. 2.23 Morgan's "safety hood"

2.2 The Morality of Military Engineering

The above engineers were all in some way and at least during some part of their careers engaged in military engineering. For the most part, these military engineers were not morally bad people and would have insisted that working for the military is an honorable profession. This is not unexpected. Sociological studies have found that all employees, including engineers, have a tendency to focus on the task at hand and to bracket this from their personal moral values. Work seems to be a special case, so to speak, and one's personal values do not apply. This compartmentalization allows employees to function at a level of minimally acceptable ethical behavior [1].

One might be tempted to argue that weapons research and military engineering are *always* morally wrong because both killing and enabling killing are blameworthy under all conditions. But it is not all that simple, and it is possible to mount strong arguments for doing military engineering. Most moral systems agree that killing in self-defense is not morally wrong and can be justified under certain special circumstances. Killing as a means of preventing or stopping the abuse or the certain death of innocents is also ambiguous. It is not possible, therefore, to categorically declare military engineering morally bankrupt.

When military engineers are asked to defend their work on moral grounds, they often use one of four ethical arguments:

1. Some argue that military research and armament production is a gift from God, and therefore "it is inherently good" and thus morally acceptable. Many people in the military have strong religious convictions and believe that their cause is just and that God is on their side. Given such a belief, it is not difficult to assert that military research is also a gift from God and, therefore, inherently good. But if military research is a gift from God, and therefore finds favor in God's eyes, then everything else must also be a gift from God. It is not possible to know what is and what is not a gift from God. Is it possible that racism, religious intolerance, and terrorism are also gifts from God? If so, does that make them either admirable or morally acceptable? [28].
2. The second argument states that working in military research and development is a civic duty. Since the country as a whole has agreed that this activity is to continue, and since the people are willing to pay for it, it is the civil duty of engineers to carry out the work without worry about moral implications. This argument is hollow because it assumes that the society that hires engineers always knows what the morally right thing to do is. If we accept this argument, we would also then have to say that the cruel treatment of African Americans during the Jim Crow years was morally acceptable because the majority of the people supported it [29].
3. A third approach to arguing for armaments research and manufacturing is that the work is actually intended to promote peace. These engineers point to the winning of the Cold War with the Soviet Union as proof that a strong defense eventually results in a peaceful settlement of disputes. World history, however,

disputes this claim and instead shows that military buildups almost always result in conflicts. The Cold War was won more by the collapse of the Soviet economic system than by the superiority of our military establishment. And having this establishment available for use by political leaders has resulted in numerous local conflicts.

4. A fourth approach, credited to Samuel Florman, who wrote so eloquently in the 1970s on the value and worth of engineering and engineers, is that engineers should engage themselves in whatever needs to be done and to leave the moral questions up to those who use the engineer's skill. This is known as the "gun for hire" approach. There is no problem with doing murderous work. One is simply doing what one was trained to do, and the value in the work is the quality of the outcome. Good engineering is good engineering, regardless of how the work is eventually applied. Engineering thus is an amoral activity and engineers should not be required to make moral decisions [30]. Florman used the "responsibility of the professions" argument to press his case:

> If each person is entitled to medical care and legal representation, is it not equally important that each legitimate business entity, governmental agency, and citizen's group should have access to expert engineering advice? If so, then it follows that engineers (within the limits of conscience) will sometimes labor on behalf of causes in which they do not believe [30].

This is, however, a spurious argument. A physician is beholden to give care when life is threatened but is not at all required to perform surgery or prescribe medications if there is no need for it. The lawyer may defend a client that he or she believes is guilty, but can always ask to be relieved of the duty. Similarly, there is no moral responsibility for an engineer to perform engineering on a project that he or she believes to be morally reprehensible.

As demonstrated by some of the short biographies above, clearly some engineering is not moral, even if it is good engineering. Engineers do have an opportunity and a professional mandate to make moral decisions. Stephen Unger suggests that engineers should

> … endeavor to direct their professional skills toward conscientiously chosen ends they deem, on balance, to be of positive value to humanity, declining to use those skills for purposes they consider, on balance, to conflict with their moral values [31].

But this mandate is not without problems. Debra Johnson, for example, has pointed out that some engineers might have some pretty rotten moral values. Should this absolve them of any blame for unethical behavior? Secondly, most engineers who share societal moral values might not realize that their decisions are morally laden. Asking an engineer to do what he or she thinks is the right thing to do therefore opens up some serious difficulties and leaves the public no better off [32].

Unger replies that engineers share a core set of values such as truthfulness and openness, and that the vast majority of engineers respect human rights and human welfare. He argues that we achieve the greater good when we ask engineers to decline work that might be in conflict with their moral values [33].

John Forge points out that engineers have both a "positive duty" and a "negative duty" to do the right thing. The former is the duty to do well, while the latter is the duty to prevent harm. For example, during the Second World War when Britain stood alone against the Nazi war machine, did engineers have a "positive duty" to work on military research in order to prevent Germany from invading England? Their work on radar and in deciphering the German Enigma code prevented harm to their countrymen and saved England from being invaded by Nazi Germany. Was it also a "positive duty" for American nuclear scientists to manufacture an atomic bomb? Forge continues:

> If we propose that the injunction "Never do war research" be included among the responsibilities of the engineer, and if this is understood as absolute, then we have to deal with the objection that, in times of extraordinary emergency, the engineer should be free to do weapons research. So must we treat the injunction as one that could be over-ridden in extraordinary circumstances? And can it in any case be defended as the norm? In peacetime, when there are no enemies bent on conquest and genocide, can the injunction be imposed? I would answer "yes" to all these questions [33].

One approach used to judge the level of immorality of military work is to argue that one should only work on defensive projects. The department, after all, is not the Department of War (which it used to be until the 1930s) but the Department of Defense, and thus much of the work they sponsor has to be defensive. The argument is that engineers who have qualms about doing military work should limit what they do to defensive armament systems, that this will be morally permissible. But the trouble is that it is difficult if not impossible to classify a weapon as defensive or offensive. Perhaps there are some that can be so cataloged, but most cannot. From the simplest weapon, the rifle, for example, to the most intricate, the laser-guided smart bomb, these are all both offensive and defensive. In some cases one might argue that body armor, or armor for the trucks in Iraq, is purely defensive, but the point is that by having such armor or protection, the soldier or the truck becomes capable of undertaking clearly offensive actions.

All weapons are therefore both offensive and defensive, depending on one's perspective. For example, the Star Wars system was viewed by the Reagan administration (and subsequent administrations) as purely defensive, but Russia believes that it is an offensive weapon that destabilizes the armaments balance. Russia's view is that if the shield is indeed successful, the USA would be free to use its nuclear arsenal without fear of retribution. Also, the mere presence and availability of weapons and an army make some leaders more likely to use them for nondefensive purposes. We have witnessed the use of our forces, which are in place to defend against aggression, as an offensive weapon in Iraq.

Moving along a continuum – from military work without concern for consequences to work limited by one's moral concerns – we come to the option that an engineer should not do any military work under any circumstances and instead endeavor to use his or her skills for the promotion of peace. This position, which in effect argues for pacifist engineering, is the limiting position. As Albert Camus, in his powerful booklet "Neither Victims Nor Executioners," writes:

> All I ask in the midst of a murderous world, we agree to reflect on murder and to make a choice. After that, we can distinguish those who accept the consequences of being murderers themselves or the accomplices to murderers, and those who refuse to do so with all their force and being [34].

Such is not an easy decision and may not even be the best one, all things considered. There may be times and instances when doing military work will be the most morally acceptable alternative. Perhaps the wisdom of Aristotle, seeking moderation in all things, is again the best alternative.

2.3 Military Work in the Legal and Medical Professions

Engineering is not the only profession that wrestles with the application of professional skills to the military and the making of war. Consider the military lawyers who have justified the need for the concentration camp called Guantanamo in Cuba. Should an attorney condone the torture of prisoners, knowing that this is both illegal and immoral?

Some attorneys assigned to prosecute the prisoners have already rebelled against the mistreatment received by these prisoners. Writing in the *Washington Post,* former military prosecutor Darrel Vandeveld says that prosecutors were bullied by Bush political appointees to bring charges before the cases were ready. He also described how one 16-year-old detainee was "hooded, slapped repeatedly across the face and then thrown down at least one flight of stairs." Detainees were also subjected to sleep deprivation and other forms of abuse. Finally Vanderveld decided to quit his post, saying he had ethical qualms about the system. "I am ashamed that it took me so long to recognize the stain of Guantanamo, not simply on America's standing in the world, but as part, now, of a history we cannot undo," Vandeveld writes [35]. The American Bar Association has not taken a stand on the treatment of prisoners at Guantanamo and apparently does not intend to do so.

Another case of professionals being confronted by ethical dilemmas is the use of medicine in warfare. Historically, conflicting armies have used physical means to kill each other. Chemical warfare in modern times was first used by the Germans in the First World War and by Saddam Hussein against the Kurds in Iraq. The use of chemicals is theoretically not permitted according to the Geneva Convention, but stocks of chemical warfare agents still exist in arsenals of all major powers, including the USA.

A third method of subduing the enemy is to use biological agents such as anthrax or other lethal viruses and microorganisms. Even more insidious is the use of medicines that were developed to help those who suffer from various illnesses. For example, what would happen if the drug Valium were somehow introduced into the bodies of an opposing army? The effect would be relaxation, lethargy, and inability or unwillingness to fight. Valium, if it could act quickly enough, could also be used as a means of riot control by law enforcement officials.

This use of "drugs as weapons" has worried many healthcare professionals and prompted the Royal Academy of Medicine in the UK to convene a special study committee. Their conclusion was that the use of a drug as a method of warfare would constitute a violation of the 1925 Geneva Protocol and the 1993 Chemical Weapons Convention (CWC). "The committee believed that the use of drugs as weapons is simply not feasible without causing great harm to non-combatants. They believed that it would be impossible to deliver the exact dose of a drug to the right people, thus causing unacceptable collateral damage." The bottom line, according to the British Medical Association, is that any physician who participates in a "drugs as weapons" program is acting unethically and can be censured [36].

If physicians can declare the use of their skills in warfare morally unacceptable, what would it take for the engineering profession to do likewise?

2.4 Military Engineering at American Universities

It seems reasonable that if peace engineering is to prosper and someday take its place alongside other engineering endeavors, then we need to engage American universities. At the present time, the situation at engineering schools is rigged in favor of military engineering. Zussman writes:

> Because engineers are embedded in industry or the military, they typically serve the ends of profit-making, or defense, and it has not traditionally been considered a professional duty to question those ends [37].

The funding of graduate students by the military has long-term effects. Students become experts in a specific area, and if this is of importance to the military, the students, upon graduation, have to seek funding from the military to continue their work. They would have to completely retool their skills to avoid this, and thus the military can control their academic careers [38].

The military needs engineering talent, and they thus do aggressive recruiting. Manion and Kam write:

> One only needs to look at the history of the engineering profession to see how closely engineering schools and large corporations work together to tailor an engineering curriculum suited to the immediate needs of the military-industrial complex [39].

American universities, now so critically and financially in need of the overhead dollars that come from doing research, can only push for more funding, and the defense establishment is the governmental sector with the most money. As an example of the kind of slithering beggars university presidents can become when they need more money, we have the testimony of the president of the Association of American Universities, an elite group that admits only 50 of the largest and most prestigious universities. The following testimony in front of the House of Representatives Committee on Appropriations shows how students are used as pawns to squeeze more money out of the government:

> I think it … would be helpful to the Department of Defense to enlist the loyalty of a group of students with funds that it awards, to enable students to pursue their graduate studies with the sense that they are beginning an engagement in work that is of interest, that will ultimately be of interest to the Department of Defense and related agencies.
> So I think from that point of view, to the extent that the Defense Department supports graduate students in the sciences and engineering, it is beginning to build cadres of scientists and engineers who will be participants in the programs in the future …
> … If they are engaged early in work that is intellectually stimulating to them and that has some promise for the future and is supported by the Department of Defense, it seems to me you are well on the way to having them hooked into that enterprise for a long time [40].

"Hooked"? The presidents of our universities want to "hook" our students into doing military research?

Some years back the Office of Naval Research was casting around to find a suitable university where it could set up shop and condone high-level biological research. It eventually settled on MIT and offered the university huge research funding. The faculty of the Department of Biology were coerced into accepting this offer, but not without having their say first. Professor Jonathan King, the head of the department, wrote an article that described the process:

> Funding for biological research by the military serves several purposes. It contributes to the increasing incorporation of the university into the military-industrial complex. It provides a veneer of respectability to cover the support of the military for its more destructive projects. It increasingly focuses academic research on problems of concern to the military. And it provides direct and indirect support for the resurgence of biological warfare research [41].

Some universities, such as the American Military University, have been established to serve the needs of military personnel, offering online degrees for returning servicemen and women. Their Web site advertisement reads:

> AMU students are active, working adults in the military, public safety and national security sectors, and beyond. Founded to provide relevant and affordable education to the military, AMU today serves a variety of students seeking liberal arts and professional studies degrees, and we place special program emphasis on programs in homeland security, national security, intelligence, and emergency and disaster management. Everyone associated with the University shares a passion for providing curriculum, class delivery, and service uniquely designed to meet the needs of those who serve others in their communities.

The high percentage of engineers engaged in defense-related industry is not altogether a random selection. The armed forces go to a great deal of trouble and expense to recruit engineers, often with advertising that plays up the exciting nature of military engineering and suggests that alternatives are boring and worthless. Figure 2.24 is an example of this advertising. The ad suggests that the only cool thing for graduating engineers to do is to work for the military. It is no wonder that students perceive that the most excitement and the best jobs are all in the military/industrial complex.

In fact, when Dwight Eisenhower warned us of the military/industrial complex, he should perhaps have called it a *military/industrial/university* complex.

Fig. 2.24 A recruitment ad by the US Navy directed at engineering students

References

[1] Bowen WR (2006) Prioritising people – outline of an aspirational engineering ethic. In: Proceedings of the workshop on engineering and philosophy, Delft, The Netherlands
[2] Langley C (2005) Soldiers in the laboratory: military involvement in science and technology and some alternatives, Scientists for Global Responsibility, Kent, UK
[3] Stockholm International Peace Research Institute (2003) Yearbook. Stockholm
[4] Martel WC (2001) The technological arsenal: emerging defense capabilities. Smithsonian Institution, Washington, DC
[5] Feldman J (1989) Economic conversion: an alternative to military dependency in the university. In: Mitcham C, Siekevitz P (eds) Ethical issues associated with scientific and technological research for the military. New York Academy of Sciences, New York
[6] Papadoupoulos C, Hable A (2008) Including questions of military and defense technology in engineering ethics education. In: Proceedings of the ASEE conference and exposition
[7] Singer P (2009) Wired for war. Penguin, New York
[8] Michener JA (1984) Tales from the South Pacific. Fawcett, New York
[9] Kajencki FC (1998) Military engineer of the American Revolution. Southwest Polonia, Lafayette, LA
[10] Smith JE (1984) Dictionary of American military biography, vols 77–80/81–84 'Lucius Clay.' Gale Research, Farmington Hills, MI, pp. 685, 686
[11] Katz SM (2002) The hunt for the engineer: the inside story of how Israel's counterterrorist forces tracked and killed the Hamas master bomber. Lyons, Guilford, CT
[12] Prüfer K (1946) Testimony of crematorium engineers. Jewish Virtual Library: *www.jewishvirtuallibrary.org/jsource/Holocaust/crematorium_test.html*
[13] Redford J (1992) http://world.std.com/~jlr/doom/bull.htm
[14] Lowther W (1991) Arms and the man - Dr. Gerald Bull, Iraq, and the supergun. Presidio, New York
[15] William L (1991) Iraq and the supergun: Gerald Bull: the true story of Saddam Hussein's Dr. Doom. Macmillan, London
[16] Drake S (1978) Galileo at work: his scientific biography. University of Chicago Press, Chicago
[17] Roland A (1989) Hephaestus and history. In: Mitcham C, Siekevitz P (eds) Ethical issues associated with scientific and technological research for the military. New York Academy of Sciences, New York
[18] http://en.wikipedia.org/wiki/Strategic_Defense_Initiative
[19] Lipton E (2008) Pork-barrel insider drained missile-defense millions. New York Times 158(54,461):1. 12 October 2008
[20] Devreese JT, Vanden Berghe G (2008)'Magic is no magic': the wonderful world of Simon Stevin. WIT Press, Boston
[21] Langewiesche W (2005) The wrath of Khan. The Atlantic, November 2005
[22] O'Brien JM (2007) Killer robots could replace soldiers. Fortune, 4 December 2007
[23] Ratliff E (2009) Shoot!. New Yorker, 23 February 2009
[24] http://ethics.calpoly.edu/pr_022309.html
[25] Dinger's Aviation Page, http://freespace.virgin.net/john.dell/index.htm
[26] Mitcham C (1989) The spectrum of ethical issues associated with the military support of science and technology. In: Mitcham C, Siekevitz P (eds) Ethical issues associated with scientific and technological research for the military. Annals of the New York Academy of Sciences, vol 577. New York Academy of Sciences, New York
[27] Zachary GP (1999) Endless frontier: Vannevar Bush, engineer of the American century. MIT Press, Cambridge, MA
[28] Dinegar RH (1989) The moral arguments for military research. In: Mitcham C, Siekevitz P (eds) Ethical issues associated with scientific and technological research for the military. New York Academy of Sciences, New York

[29] Kemp K (1989) Conducting scientific research for the military as a civil duty. In: Mitcham C, Siekevitz P (eds) Ethical issues associated with scientific and technological research for the military. New York Academy of Sciences

[30] Florman S (1970) Moral blueprints: on regulating the ethics of engineers. Harper's, October 1970

[31] Unger S (1982) Controlling technology: ethics and the responsible engineer. Holt, Reinhart and Winston, New York

[32] Johnson D (1989) The social/professional responsibility of engineers. In: Mitcham C, Siekevitz P (eds) Ethical issues associated with scientific and technological research for the military. New York Academy of Sciences, New York

[33] Forge J (2006) The morality of weapons research. In: Vesilind PA (ed) Peace engineering. Lakeshore Press, Woodsville, NH

[34] Camus A (2002) Neither victims nor executioners. In: The power of nonviolence. Beacon Press, Boston

[35] Weiss C () How a military prosecutor changed from gung ho terrorism fighter to Gitmo critic. ABA J: *http://abajournal.com/news*

[36] Board of Science (2008) The use of drugs as weapons: the concerns and responsibilities of healthcare professionals. British Medical Association, London

[37] Zussman R (1985) Mechanics of the middle class: work and politics among American engineers. University of California Press, Berkeley

[38] Roth B (1989) The moral arguments against military research. In: Mitcham C, Siekevitz P (eds) Ethical issues associated with scientific and technological research for the military. New York Academy of Sciences, New York

[39] Manion M, Kam M (2000) Engineering ethics at Drexel University. In: Proceedings of the American Society for Engineering Education annual conference

[40] Rosenzweig R (1989) Congressional testimony on 15 May 1985, Department of Defense Appropriations for 1986, Committee on Appropriations, 99th Congress. US Government Printing Office, Washington, DC; quoted in B. Roth, Moral arguments against military research. In: Mitcham C, Siekevitz P (eds) Ethical issues associated with scientific and technological research for the military. New York Academy of Sciences, New York

[41] King J (1989) Biology goes to war. Sci People 20(1), 1988; quoted in: Mitcham C, Siekevitz P (eds) Ethical issues associated with scientific and technological research for the military. New York Academy of Sciences, New York

Chapter 3
Civilian Engineering

Not all engineering is military engineering, and not all engineers are engaged in military work. The tradition of engineering in the civilian sector has also been important in the development of human civilization. The line between military engineering and civilian engineering is often not clear and not every engineering project or activity can readily be placed in one or the other of these categories. Nevertheless, some engineers work on projects that mostly benefit the civilian sector. The exemplars below, again listed at random, are useful in defining what we mean by civilian engineering.

The stories are of two kinds – those where the engineers chose to take seriously their commitment to society, and those where other values caused them to forget or ignore the first canon of the code of ethics. What can we learn from the stories of those who took the high moral road and those who let down the profession and, ultimately, themselves?

3.1 Exemplars of Civilian Engineering

One of the benefits of reading about the lives of engineers is that it is easy to insert oneself into their jobs and to venerate those engineers who have done it right, but at the same time to imagine what pressures they must have experienced during their careers that might have led them down the wrong path. Most of the engineering exemplars introduced in this chapter had honorable careers, and we look up to them with respect and admiration. But not all engineers are able to navigate through their careers without serious moral stumbles.

Rushworth Kidder, in his book *Moral Courage*, calls truly bad ethical decisions "career-ending moves" [1]. He illustrates this with a true story of the CEO of the Bath Iron Works who was bidding on a Navy contract and discovered that the

P.A. Vesilind, *Engineering Peace and Justice*,

Navy personnel with whom they had been talking had inadvertently left behind a copy of the proposal from Bath Iron Works' competitor. What to do? If they looked in the proposal, they would know what the competitor was proposing and could write their own proposal in a way that would assure them of winning the contract. Or, they could leave the document unopened and return it to the Navy procurement officer. The CEO chose to do the former. When people in his own firm balked at looking in the Navy document, and the president of the firm found out about it, the CEO was fired. It was his "career-ending move."

John Smeaton (1724–1792)

Fig. 3.1 John Smeaton

The first engineer to recognize the difference between civilian and military engineering was John Smeaton, who started calling himself a "civil" engineer. Born in 1724 in England, Smeaton showed early promise as a builder of machines and structures, and when he was finally allowed to leave his training in law, which had been imposed by his father, Smeaton became the most respected English engineer in the 18th century.

One of his famous projects was the construction of the first successful lighthouse on Eddystone Reef, south of Plymouth, a reef which lay directly in the path of a shipping channel and had been responsible for the destruction of many ships. There had been other attempts to build the lighthouse on this reef. In 1698 Henry Winstanley built a lighthouse on one of the rocks, and although it was secured by iron anchorage bars, it washed away in the hurricane of 1703. A second structure, built by John Rudyerd in 1708, was more securely attached but it was built of timber and was destroyed by fire in 1755. Smeaton's lighthouse, completed in 1759, was made entirely of interlocked Portland stone, and it took 2 years to build it, working under the most difficult of conditions in the stormy channel.

Smeaton went on to construct many of the important bridges in England and he designed a new water supply for Edinburgh. A major achievement was the construction of the Forth and Clyde canal that crossed Scotland and provided a route for sea-going vessels between the North Sea and the Atlantic Ocean. Recognizing that his works were not of a military nature, Smeaton began to sign his name using the title "civil engineer." He and his engineering colleagues used to meet at the

Queens Head Tavern in London, and this group became the nucleus of the Institution of Civil Engineers, chartered in 1818 [2].

Margaret Ingles (1892–1971)

Fig. 3.2 Margaret Ingles

Margaret Ingels went to the University of Kentucky to study architecture but was persuaded to do mechanical engineering instead. She became the first female engineering graduate from that university when she received her Bachelor of Mechanical Engineering in 1916. She was also the second woman engineering graduate in the USA and the first woman to receive the master's degree in mechanical engineering.

Following graduation she went to work for the Carrier Engineering Corporation, where her interest in "conditioned air" began. She then joined the American Society of Heating and Ventilating Engineers research lab where she studied air conditioning for 6 years. In 1931 she returned to Carrier in Syracuse, NY, and remained there until her retirement.

Ingles became a strong and effective spokesperson for the engineering professions, and especially for women who wanted to study engineering. She gave more than 200 speeches all over the country. In her most famous speech, "Petticoats and Slide Rules," which was first presented to the Western Society of Engineers in 1952, she gave credit to the women who had preceded her in engineering.

> The woman who joins the profession of engineers today, tomorrow, and tomorrow benefits by a rich heritage bequeathed to her by [those who came before]. She assumes automatically the responsibility to further prove that petticoats and slide rules are compatible, and she must not carry the responsibility lightly. Her task is to widen the trails blazed for her – and more. She must build them into great highways for women engineers of the future to travel, free of prejudices and discrimination.

In 1957 she received an Honorary Doctor of Law degree from the University of Kentucky, and a new dormitory was named Ingles Hall in her honor [3].

Peter Palchinsky (1875– 1929)

Fig. 3.3 Peter Palchinsky shown here with his wife Nina

Peter Palchinsky's greatest problem was that he took seriously the idea that engineers should hold paramount the health, safety, and welfare of the public.

Born to poor parents in 1875 in central Russia, he worked his way through school, graduating with a degree in mining engineering from the prestigious institute at St. Petersburg. On leaving school, he was hired to investigate the causes for the low production of coal at the tsar's mines in the east. He found that the miners were living and working under appalling conditions, with no concern for their occupational safety. His report, at first applauded by the leaders in St. Petersburg, got Palchinsky into trouble once the implications of what he was suggesting were realized. Providing the miners better working conditions would have caused great upheaval in other parts of the frail economy. Palchinsky was soon arrested for speaking the truth and spent some time in jail. He escaped to the West and developed a distinguished career as an engineer designing harbors and port facilities, always recognizing that these facilities were large systems that included the need for worker protection and comfort.

Although he had a thriving engineering practice, he missed Russia, and, after being pardoned by the tsar, in 1913 he went back. When the Bolsheviks came to power in 1917, Palchinsky was arrested once again as a collaborator and spent more time in jail. His honesty and skills prevailed, however, and he eventually became a well-respected engineer, founding a journal and an institute to study mining engineering. But he made the mistake of not following the party line and was highly critical of the top-down planning, the emphasis on huge projects that neglected secondary consequences, and the poor living conditions of the workers. In a 1927 report on Magnitogorsk (the "Steel City") Palchinsky concluded that the project had failed because there was no water transportation, no coal nearby, no local labor force, and no idea how much iron ore was available. The workers had been promised a "garden city" to live in but instead got barracks with open sewers, downwind from the blast furnaces. Because of labor shortages, 30,000 prisoners were used, fully 10% of whom died during the first winter. Stalin would not tolerate people who spoke truthfully about his projects and finally had Palchinsky arrested and shot.

His wife Nina (shown in the above photograph with Peter) learned her husband's fate from a small news article describing the crimes against the revolution supposedly committed by Peter Palchinsky. As the wife of a criminal, she was also suspect and was eventually sent off to the labor camps so graphically described by Alexander Solzhenitsyn.

In the *Ghost of the Executed Engineer*, an exceptional book describing the conditions of engineering in the former Soviet Union, Loren Graham argues that the destruction of the engineering profession in Soviet Russia led directly to the downfall of the USSR [4]. Emasculation of the engineers eliminated the one group of educated persons who could point out the obvious stupidity of the projects concocted by the state planners.

William LeMessurier (1926–2007)

Fig. 3.4 William LeMessurier

With a degree from the Harvard Graduate School of Design and a master's degree from MIT, LeMessurier built a well-regarded engineering firm that specialized in high-profile projects. In 1978 he was asked to design a new skyscraper that Citicorp wanted to use as its New York headquarters.

The design called for an attractive, functional, and imaginative 59-story building. Because of space and light restrictions, the architects designed a building that seemed to float on four columns nine stories high, providing light and space below and enhancing the visual appearance of a new church building on the corner of the lot. To achieve this, the architects suggested that the four columns be placed in the middle of each side instead of at the corners. LeMessurier decided to use a unique form of construction, transferring forces to the four columns by means of V-shaped beams.

Engineering design is trial and error. A structure is postulated, and the loads on that structure are then estimated. Using mathematical principles and well-tested equations, the effect of these loads on the structure are calculated. In the case of the Citicorp building, LeMessurier's engineers calculated, in addition to other live loads, the effect of wind and decided that with a damper mechanism in the attic of the building the building would be able to withstand high winds.

The Citicorp building was constructed and occupied, and the client was very pleased with the result. Then out of the blue LeMessurier got a telephone call from a student who told LeMessurier about a homework assignment she had done. The student had calculated the ability of the building to withstand wind loads and found that as long as the winds were from the side of the building, the structure seemed to be secure, but if the winds hit the building at its corners, so-called quartering winds, it would be possible to topple the building when subjected to moderate winds.

LeMessurier told the student that, in effect, she didn't know what she was talking about. But the call got him to think about the wind loads, and he redid some

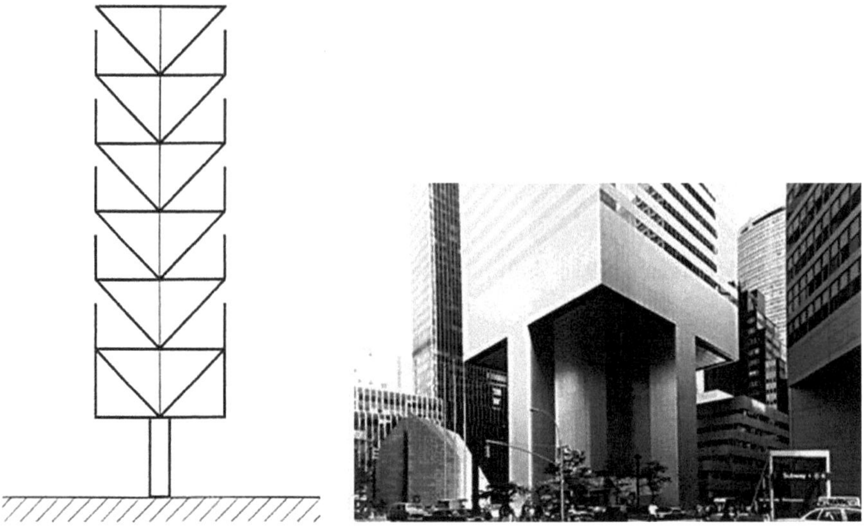

Fig. 3.5 The structural system of the CitiCorp building

calculations. To his surprise, it seemed that the effects of the 45-degree winds were much greater than he had originally estimated.

He became really concerned when he remembered that the method of structural construction had been changed during the erection procedure. Instead of welding the joints, the construction engineers had substituted a newer standard using bolted joints. If the effect of these joints was included in the analysis, it became painfully clear that, should the building be hit by quartering winds, some of the beams in the building would not be able to withstand the live load and the building would topple. Weather records showed that such winds might occur once every 16 years.

That probability was unacceptable and the risk was far too high. Should the building fall, thousands of people would die. The only person in the world who knew that was LeMessurier. He contemplated his options and said, “Thank you Lord for making this problem so sharply defined that there’s no choice to make” [5].

With the consent of the owners, he hired disaster engineers who planned for evacuation should a storm arise. Next he instrumented the entire building with strain gauges and set a 24/7 watch on the damper mechanism in the top floor to make sure it functioned perfectly. Then LeMessurier started to strengthen each of the V-shaped joints where bolts had been used by welding in supporting plates. Since the structural members were all inside the building, he could do all the construction from the inside, thus avoiding embarrassing questions and possible panic. Within months, all of the joints had been strengthened and the building became one of the safest in New York, able to withstand the highest winds that could reasonably be expected to occur.

After discovering the problem, LeMessurier could have done nothing, believing that by revealing this information he would have lost stature and respect in the engineering community and hoping that the series of events that accumulatively would have led to a catastrophe would not occur within his lifetime. Instead, he chose the honorable alternative, perhaps remembering the engineering code of ethics: "The engineer shall place paramount the health, safety, and welfare of the public." The risk was so great, both in terms of its probability and its magnitude, that there was little choice for him or for the owners of the building.

As it was, by conducting himself in such an honorable manner, he actually gained considerable stature in the profession and in the public's view. After the "fifty-nine-story crisis" was resolved, he made himself widely available as a lecturer in colleges and universities, always speaking with candor about what could have been the greatest disaster in his otherwise illustrious career, but which turned out to be his greatest engineering triumph. He never recorded the name of the student who had alerted him to the problem, and her name remains a mystery.

Les Robertson (1928–)

Fig. 3.6 Les Robertson

Being at the top of the profession often affords opportunities to do the most imaginative engineering, and this was true for Les Robertson. Educated at the University of California at Berkeley as a civil engineer, Robertson went on to work with some of the best structural consulting firms in the USA, eventually establishing his own firm, Leslie E. Robertson Associates. In the 1960s, while employed by the firm headed by John Skilling, he became the lead design engineer for the New York Port Authority World Trade Center buildings.

Architect Minoru Yamasaki had designed the two towers, and now it was up to Robertson and his team to make the design work. Robertson used the brilliant concept of a central core and external cross-bracing. The structural system included a prefabricated building façade, with columns acting as wind bracing. Placing the bracing on the outside opened up the interior, resulting in large office spaces without columns. The floors were of prefabricated trussed steel that also acted as a diaphragm to stiffen the outside wall against lateral buckling forces from wind-load pressures. A major innovation was the elevator system, which allowed high-speed cars to ride part way up, thus reducing the size of the shafts and the air pressures created by the moving cars. It was an innovative and effective design and earned Robertson many accolades.

Fig. 3.7 The structural system for the New York Port Authority Twin Towers

Then, on September 11, 2001, fully loaded Boeing 757s slammed into each building, causing both to collapse.

After a thorough investigation, it was determined that the most likely cause of the collapse was the intense and prolonged heat generated by the fires. Although the temperature would have been well below the melting point of steel, it was high enough to cause steel to become soft, allowing for plastic deformation. Most engineers believe that the connectors that held the floors were the first to go, and as one floor collapsed into the one below, the kinetic force was sufficient to cause sequential collapse. By the time the collapsing floors in the section on fire reached the floors that were not on fire, the kinetic forces were many times higher than anything that they could support, and all of the floors pancaked.

The question on everyone's mind was: "How was this possible?" Why did the designers not anticipate this "load" and design for it? The answer is that at the time the towers were built, the largest airplane flying was a Boeing 707, and the most likely scenario was that the towers would be hit by a plane lost in the fog trying to land, thus lightly loaded with fuel and flying at a slow speed. The two planes that slammed into the towers were larger and heavier, with full fuel tanks, and traveling at about 850 km/h. Nothing Robertson could have designed would have withstood this assault.

Did Robertson make a mistake by underestimating the possibility of what eventually occurred? Why did he not run a risk analysis where a much larger airplane, fully loaded with jet fuel, would crash into the buildings? If he had, the buildings probably would not have been built. But would this have been a reasonable risk? How much risk should civil engineers design for? If the answer is zero risk, then nothing would ever be built.

Elijah McCoy (1843–1929)

Fig. 3.8 Elijah McCoy

Elijah McCoy was born in Colchester, Ontario, Canada, in 1843 to George and Mildred McCoy, who were escaped slaves, having fled from Kentucky to Canada, "riding" the Underground Railroad. After the Civil War, McCoy and his family moved to Ypsilanti, Michigan, where his father worked in the logging industry. They saved enough money to send young Elijah to study engineering in Scotland, but on his return to the USA, he found no work as an engineer and eventually became a fireman on a steam engine. His job included oiling the wheels of the engine, and his inventive mind went to work figuring out how to lubricate moving parts while they were still in motion. In 1872 his invention for lubricating steam engines was awarded a patent, and with the help of investors, McCoy went into business. The lubricating device was so effective that machinery manufacturers and users demanded that their machines be equipped with the McCoy lubricators and would not accept alternatives. Thus comes the old saying that the best is "the real McCoy."

Perhaps the irony of this story is that McCoy's commitment to society was far greater than the commitment society made to him in return. He had to work under restrictive and adverse conditions, and yet his skill and inventiveness prevailed.

George Waring, Jr. (1833–1898)

Fig. 3.9 George Waring

George E. Waring, Jr. was educated at College Hill, Poughkeepsie, NY, and then studied agriculture with a private tutor. In 1857 he became the agricultural and drainage engineer of Central Park in New York City. With the beginning of the Civil War, Waring received a commission in the US army, was involved in numerous battles, and eventually rose to the rank of colonel in the cavalry. After the war he settled in Newport, RI, and became the manager of a large farm, but this was too easy and he eventually became a full-time engineering consultant.

After the destruction and deprivation of the Civil War, many southern cities were poor and unsanitary. In Memphis, TN, death rates from communicable diseases were so high that this problem caught the attention of the nation and a commission was formed to study the city's health problems. Waring proposed to construct a sewerage system limited only to household wastewater instead of a combined system that would carry both storm water and human waste. His plan, which he based on the small-diameter sewers first promoted by Edwin Chadwick in London, was expected to be only one tenth the cost of constructing a combined system. After much political fighting, the city decided to adopt Waring's plan, and a system was constructed consisting of 6-in. vitrified clay pipes with flush tanks leading from homes into increasingly larger collecting sewers.

Waring believed that these sewers were necessary in order to enhance public health, but his reasoning was faulty. He believed in the eventually discredited miasma theory of disease – that people became ill because they came into contact with sewer gas. A totally buried and tight system was supposed to reduce the incidence of cholera, typhoid, and other such diseases because these systems did not allow the miasma to waft into the community.

After the sewers had been constructed the incidence of communicable diseases dropped markedly and Waring claimed the system to be a success. However, most people considered the system a failure because of operational problems. The small lines from the households often clogged and had to be cleaned with snakes, and when the collecting lines clogged, streets had to be dug up to unblock the sewers. Eventually manholes were constructed which, if factored into the original cost, would have significantly increased the price of the system.

Nevertheless, the controversy as to whether a city should build separate sewers instead of combined sewers raged for decades, with most engineers favoring combined systems since they were less expensive to build than two separate systems. At that time there were no wastewater treatment plants, and all water – storm and wastewater – went to the same convenient place, such as the Mississippi River in the case of Memphis, and so the engineers had a good point. Stormwater, however, has only marginal polluting potential while sanitary wastes (sewage) is highly polluting. Building wastewater treatment plants for a mixture of stormwater and sewage is exorbitantly expensive. Only by separating the sanitary wastes from stormwater is it economically possible to build wastewater treatment plants. Cities that did not build separate sewers ended up spending a great deal of money separating them later. George Waring had been right all along about the sewers, but dreadfully wrong about how disease was transmitted.

Thomas Midgley, Jr. (1889–1944)

Fig. 3.10 Thomas Midgley, Jr.

Thomas Midgley was born in Beaver Falls, PA and went to Cornell University to study engineering. After graduation he worked for Delco, where one of his first assignments was to figure out a way of eliminating the loud knocking that plagued internal combustion engines. The knocking was destructive, and as engines were continually improved by using higher engine-compression ratios, the knock problem worsened. Up to that time it was assumed that the problem was with the engine, but Midgley demonstrated that the real culprit was the fuel. The gasoline used in those days exploded very rapidly in the pistons, causing the cylinders to be jammed into the crankshaft. Midgley discovered that if tetraethyl lead was added to the gasoline, the presence of the lead slowed down the explosion, pushing the cylinder smoothly into the piston and eliminating knock.

But there was a problem. The lead in the gasoline coated the valves and eventually caused the engine to stop. Midgley found that if dibromide was added to the gasoline, the lead would not form deposits, and the engine would run smoothly. In 1923, "ethyl gasoline" became the fuel of choice for all cars and trucks and powered the automobile industry. Midgley's discovery also had a military use. With the knocking problem solved, airplanes could use more powerful engines, increasing both their speed and range.

Looking back at the discovery, tetraethyl lead was not an obvious choice for a fuel additive. The benefits of lead in gasoline were discovered as far back as 1854, but the public use of leaded gasoline had been rejected because of its toxicity. It was well known that prolonged contact with the chemical can cause hallucinations, difficulty in breathing, and, in the worst cases, madness, spasms, palsy, asphyxiation, and death. After Midgley's discovery, the US government approved the additive for gasoline only because lead was at very low concentrations and because the company promised to use great care in the manufacture of leaded gas. The effect of lead on public health from automobile exhaust was apparently not perceived as a problem.

Lead poisoning from inorganic lead had been known since the 18th century. Even Benjamin Franklin, a printer who set lead type, had warned his workers of the dangers of lead. But Midgley did not take enough precautions and developed lead poisoning as a result of contact with tetraethyl lead. He had to take time off from work to recover, but then, excited by the usefulness of ethyl gasoline, he became a spokesman for the additive and ignored the health issues. Only in his later years did he reconsider and start to question the safety of lead in gasoline [6].

In the 1930s, Midgley was charged with finding an inexpensive, nontoxic refrigerant for use in household appliances. Up to that time, the refrigerants used, such as carbon dioxide and ammonia, were either terribly inefficient or toxic. Midgley discovered that a class of manufactured organic chemicals (not found in nature) called chlorofluorocarbons, or CFCs, were ideal for that purpose, and Freon was born. Freon did not burn and did not biodegrade, so it did not have to be replaced in compressors. Most importantly, it was relatively inexpensive to manufacture. Always the showman, Midgley demonstrated the nontoxicity and nonflammability of Freon at a widely publicized press conference by filling his lungs with it and then blowing out a candle [7].

The catastrophic global effects of CFCs were not discovered until the 1970s when Mario Molina and Sherwood Rowland at the University of California at Irvine realized that the presence of CFCs in the stratosphere would have profound environmental consequences. Other scientists had recognized the effect of chlorine on stratospheric ozone, but none had shown that CFCs would have a dramatic impact on ozone concentration. Molina and Rowland published their findings in *Nature* in 1974 and immediately became targets of severe scientific criticism from industrial interests. The two scientists persevered, however, and went to great lengths to publicize their results and to testify at congressional hearings. Finally, scientists at DuPont, the largest manufacturer of CFCs, acknowledged that Molina and Rowland were correct and pledged to cease manufacture of these compounds. (The fact that DuPont had already developed an alternative refrigerant no doubt played a role in their decision.)

Midgley has the distinction of having invented two of the most environmentally destructive chemical agents ever produced by humans – leaded gasoline and CFCs. And yet George Midgley was an honorable man who led an exemplary professional life. In his mind he had done nothing remotely wrong or unethical. He responded brilliantly to the needs of society and his inventions greatly increased the quality of life, at least in the short term.

John Roebling (1806–1869)

Fig. 3.11 John Roebling

Perhaps the most famous American engineer ever was John Roebling. Born in Prussia (Germany), he graduated from the Royal Polytechnic School in Berlin, majoring in architecture and engineering. He apparently was an eclectic student, and in addition to his work in engineering he studied philosophy, becoming a particular favorite of the philosopher Hegel. After 3 years of service to the state, he immigrated to the USA and settled in western Pennsylvania, founding a small town which he called Saxonburg, after the Saxony district in Prussia. He tried farming, but he was bored and yearned to use his education for more technical purposes. His first opportunity was to work as an assistant engineer on the Beaver River, a tributary of the Ohio River, west of Pittsburgh. The overall objective was to connect the Ohio River with Lake Erie. But the era of canals soon came to an end when it became obvious that railroads could move people and goods much faster and the railroad routes were not restricted to waterways. Roebling then went to work for the state of Pennsylvania, surveying for the proposed railway across the Allegheny Mountains, connecting Pittsburgh to Philadelphia. During this time he got into the manufacture of wire rope, which he wanted to use to pull barges across the mountains between canals. The hemp rope used at that time often broke, with disastrous results. Manufacturing wire rope got him thinking about building suspension bridges. His first "bridge" was an aqueduct over the Allegheny River, which he built in record time and under adverse conditions. Next he built the Monongahela suspension bridge on the piers of an old wooden bridge that had burned down. Work in eastern Pennsylvania and Delaware prompted Roebling to move his operations east, and he built a large wire mill in Trenton, NJ.

Roebling's next project was a railway bridge across the Niagara River, which, given the chasm to be crossed, could only be done with a suspension bridge. Construction went on for 4 years, even through brutal winter weather, until 1855 when the first locomotive rumbled across the bridge. The bridge, still standing today, has two decks, one for rail and one for motorized traffic. The two decks are connected with struts, thus forming a hollow girder supported by the cables. This engineering triumph led to other jobs, including a bridge in Cincinnati spanning the Ohio River. By this time Roebling was the best known bridge builder in America.

In 1867 a group of investors in New York wanted to build a bridge across the East River to connect Brooklyn to Manhattan, and they turned to Roebling. Could it be done? The location presented a severe challenge because there were no firm shores on which to build towers. The river bottom was muddy, and nobody knew how deep the mud really was. Roebling thought it could be done if they built two piers into the mud, using caissons. He drew up plans for the bridge and started construction.

But then disaster struck. While conducting a survey on the Brooklyn side of the river his foot was crushed by a docking ferry boat. As was his custom, he refused medical aid, believing that mere willpower would solve the problem. But lockjaw set in, and Roebling died an agonizing death sixteen days later.

His son, Washington Roebling, who was one of the first graduates of the nascent engineering program at Rensselaer Polytechnic Institute, took over as the chief engineer. But he had an attack of the bends while exiting the first caisson (on the Manhattan side) and became a semi-invalid. The actual day-to-day management was taken over by his wife, Emily Roebling, who had no engineering training but who was able to communicate the wishes of Washington Roebling to the construction crews. While taking care of her husband, she studied engineering principles such as strength of materials and became knowledgeable in the essentials of structural engineering. Many of the construction workers believed that Emily Roebling was making most of the engineering decisions without even consulting her invalid husband, and that she in effect became the chief engineer. When the bridge was finally completed, they gave her the honor of being the first person to ride across the river on the newly named Brooklyn Bridge. A plaque attesting to her role in the construction of the bridge can be seen on one of the piers.

Fig. 3.12 The Brooklyn Bridge

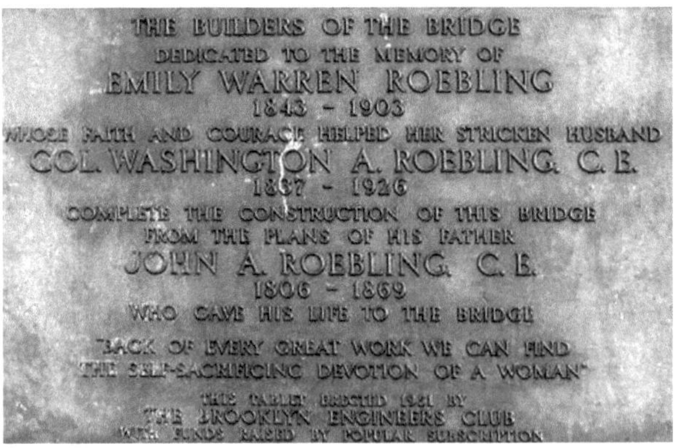

Fig. 3.13 Commemorative plaque on the Brooklyn Bridge honoring Emily Roebling

Lester Matz (1924–1985)

Fig. 3.14 Lester Matz

After working for Baltimore County for a few years, Lester Matz and an associate decided to start their own consulting company specializing in highway design. They set up an office, ordered business cards … and watched themselves go broke. No business. Matz wondered why their firm, which ought to have been competitive in terms of engineering qualifications, was not getting any work. He soon found the reason: the other firms were bribing politicians in order to get engineering work. Matz decided that he would play that same game. This was a consequential decision for Matz, since he must have known very well that what he was doing was both illegal and professionally unethical.

Matz's bribes worked and his firm was soon getting a lot of road work. He was able to hide the kickback money by paying bonuses to his employees who then "voluntarily" contributed to the fund that was used to pay the bribes. The going rate at the time in Baltimore County was a 5% kickback to the politicians, and Matz even devised a schedule of how such payments would be made. This schedule became a sort of handbook for all engineering firms and politicians in Baltimore County.

One of the first politicians he paid bribes to was the county commissioner, Spiro Agnew. Agnew was an ambitious young man who decided to run for governor of Maryland, and when he won, he became even more influential in deciding who got the lucrative road contracts. Matz, of course, was right there with the money and his firm prospered accordingly.

Then presidential candidate Richard Nixon chose Agnew to be his running mate on the Republican ticket and their victory propelled Agnew to the vice presidency of the USA. Some of the work Matz's firm was doing had been channeled to Matz when Agnew was governor, and Agnew expected his payments as previously agreed upon, even though he was now the vice president. On one occasion, Matz went to the White House with $10,000 in a plain brown envelope and gave it to Agnew in the office of the vice president. He later admitted that he came away "a shaken man," for he had just paid off the vice president of the USA.

The practice of bribery in Maryland highway construction was so pervasive that eventually the facts began to leak out, and the Baltimore County district attorney started an investigation. As the noose tightened around his neck and his own conscience got the better of him, Matz went to the authorities and confessed all. When it was all over, he lost his engineering license, was expelled from the American Society of Civil Engineers, and effectively ended his professional career. Lester Matz knew what he was doing was wrong, but he did not have the moral courage to *not* do the wrong thing.

Incidentally, Spiro Agnew was given a deal: resign from the vice presidency of the USA in return for dropping all charges. He later wrote a book in which he claimed that he had done nothing wrong, but the book revealed him to be a selfish, conniving, money-grubbing man with no principles.

If Agnew had not resigned, he would have become president of the USA when Richard Nixon was forced to resign over the Watergate debacle [8].

Dan Applegate (?–?) unknown

Fig. 3.15 Dan Applegate

The DC-3, manufactured by Douglas Aircraft, is perhaps the most storied of cargo/passenger airships. Designed to provide transport during the Second World War, it served with distinction in all theaters, dropping parachute battalions during the Normandy invasion, ferrying supplies in Burma, and even providing the backbone for the Berlin Air Lift after the war. This plane lifted Douglas Aircraft during and immediately after the Second World War to preeminence in the manufacture of cargo/passenger aircraft. After the war, building on the success of the DC-3, Douglas introduced the DC-4, a four-engine straight-wing plane with a front nose wheel, and soon followed with the similarly propeller-driven DC-5, DC-6, and DC-7. But in their enthusiasm to manufacture propeller-driven aircraft, Douglas made a strategic error. They did not anticipate the move to jet engines for commercial aircraft and allowed first deHavilland in Britain and then Boeing in Seattle to gain a foothold in the passenger airplane market. DeHavilland's ill-fated Comet kept falling out of the skies due to metal fatigue, and this left the field to Boeing, which introduced the four-engine 707, the most widely used long-distance aircraft in the world. This was soon followed by the immensely popular three-engine medium-range 727.

Douglas scrambled to catch up by introducing the DC-9, a small twin-engine craft that soon found competition from the Boeing 737. The rush to catch up with Boeing caused Douglas to experience financial difficulties. In trouble for funds, Douglas Aircraft was purchased in 1967 by the McDonnell Corporation in 1967, a manufacturer of military aircraft, and the company became known as McDonnell Douglas.

Now having the necessary capital, the new McDonnell Douglas Corporation decided to challenge Boeing in the jumbo-jet department. Boeing was ready to introduce the wide-bodied 747 and since this was a unique airplane, McDonnell Douglas decided to build the DC-10 as competition. The same idea occurred to

Lockheed Aircraft, which began work on the L1011, a craft remarkably similar to the McDonnell Douglas DC-10. Since Boeing had such a lead in the jumbo market, the race between Lockheed and McDonnell Douglas was for number two. There would be no number three.

The development costs for an airplane such as the DC-10 are immense and each corporation in effect was "betting the company" on the success of the airplanes. Although information is not available on just how much this development cost, we do know that Lockheed, after building a number of L1011s, decided that they would never make money in the jumbo-jet market and ceased to manufacture the L1011, taking a $2.5 billion loss.

The airplane as conceived by McDonnell Douglas engineers had two engines on the wings and one engine under the tail assembly. As with all large jetliners, the airplane fuselage was divided into the passenger compartment and the cargo compartment, the two separated by the passenger compartment floor. This floor was made of open trusses, providing an ideal conduit for running control lines and electric cables from the cockpit to the rear of the airplane. Three hydraulic systems, each independent of the other for the sake of redundancy, were designed and all three lines ran through the passenger compartment floor. If any two of the hydraulic systems should fail (a very low probability, so thought the engineers) the third system would still allow the pilot to fly the airplane.

New airplanes go through a thorough series of tests to prove their airworthiness, beginning with ground tests to simulate flight. Because the Federal Aviation Administration (FAA) does not have the necessary technical staff to conduct these tests, the company engineers are in effect deputized to conduct the airworthiness tests themselves, a clear conflict of interest.

In one of the DC-10 ground tests, as the fuselage was being pressurized to simulate flight, the rear cargo door flew open, causing a rapid depressurization in the cargo compartment. Since the passenger compartment was still pressurized, the sudden decompression caused the floor to collapse into the cargo compartment, severing all hydraulic and electric lines. The engineers realized that if this had occurred in flight the airplane would not have been flyable and the craft would have been lost. The investigation of this incident centered on the design of the rear cargo door and not on the placement of the control lines in the passenger compartment floor. The engineers concluded that the depressurization occurred because the door was not properly closed and that it was unlikely that this would happen again. They decided that the door was adequate but that some small modifications were needed and asked Convair, the subcontractor responsible for the door, to oversee the redesign.

Doors in the passenger compartment are known as "plug" doors because they are larger than the door opening so that the pressure in the compartment cannot blow them out. Higher pressure will just force the doors to be more tightly sealed. In the cargo compartment, however, where flight attendants cannot open the door from the inside, the doors have to be opened from the outside and cannot be plug doors.

The solution is to design a system where a manual handle on the outside of the airplane will move hooks "over center" on spools and thus pull the door shut from the inside. What the engineers did not realize is that if some object interferes with the door being shut correctly or if there is a misalignment, the hooks will not go all the way over the spools, but if the manual handle is pulled very hard, it is possible to bend the rods and stow the handle flush with the door in the shut position. The door looks like it is closed, and the warning light in the cockpit will go off even though the door is not properly closed. High pressure can then push the door out, and this is what occurred in the ground test.

The DC-10 eventually passed all its tests and was certified as airworthy. Deliveries to customers began, with American Airlines being the largest customer. In June of 1972 a lightly loaded American Airlines DC-10, taking off from Detroit, while flying over Windsor, Ontario, Canada at 12,000 feet, experienced a sudden decompression with loss of two of the three hydraulic systems. The rear cargo door flew off, collapsing the passenger compartment floor. The floor did not totally collapse, however, due to the small number of passengers, and the pilot, using the remaining backup hydraulic system and steering the plane by modulating the thrust of the engines, was able to return safely to the airport.

An investigation by the National Transportation Safety Board (NTSB) concluded that the loss of the rear cargo door caused the catastrophic decompression and ordered further modifications to the airplane to prevent similar accidents. The baggage handler responsible for shutting the rear cargo door admitted that he had difficulty shutting the cargo door and had used his knee to gain leverage before he was able to force the handle down. The NTSB concluded that the design of the rear cargo door represented a serious safety problem and that the DC-10 ought not to have been certified as airworthy. They recommended grounding all of the DC-10s then in service.

Based on this conclusion the FAA should have issued an "airworthiness directive" resulting in the grounding of all DC-10s while the modifications were being

Fig. 3.16 The DC-10, shown in this recreation as it is losing the rear cargo door. [Used with permission, Wikipedia Common, http://commons.wikimedia.org/wiki/File:Aaflight96dc10.png]

made. Such a move would have been highly detrimental to McDonnell Douglas and might have caused the company to follow Lockheed in abandoning the project. The administrator of the FAA and McDonnell Douglas management instead reached a "gentleman's agreement" to get the problem fixed as soon as possible, using the mechanism of the "service bulletin" to alert all airlines to get the aircraft modified. All doors were to have small view holes through which it was possible to see if the locking mechanism was in place. A decal with the instructions for looking through the holes was to be attached to the door. Records indicate that 2 years later not all of the existing aircraft had been so modified since the "service bulletin" is simply an advisory that can be ignored by the airlines.

After the Windsor incident, Dan Applegate, Convair's chief product engineer in charge of designing the cargo door, sent a remarkable memorandum to his superiors, warning Convair management that in his opinion the rear cargo door that had been designed by Convair engineers was unsafe. He believed that the next loss of the rear cargo door on a fully loaded DC-10 could result in total collapse of the passenger compartment floor and the loss of the airplane.

The two-page memorandum, written in a matter-of-fact engineering style, discusses the problem of the catastrophic decompression and the collapse of the passenger compartment floor, resulting in the loss of all hydraulic systems, and concludes [9]:

> … once this inherent weakness was demonstrated by the July 1970 [ground] test failure, [McDonnell Douglas] did not take immediate steps to correct it. It seems inevitable that, in the twenty years ahead of us, DC-10 cargo doors will come open and I would expect this to usually result in the loss of the airplane … It is recommended that overtures be made at the highest management level to persuade Douglas to immediately make a decision to incorporate changes in the DC-10 which will correct the fundamental cabin floor catastrophic failure mode.

Applegate's memorandum was discussed by Convair management, and he was told that if they went to McDonnell Douglas with these concerns, Convair would be vulnerable to criticism and possible liability. Convair management believed that since the aircraft was certified by the FAA, Convair should not get involved. In effect Applegate was told to shut up, which he did.

In March of 1974 a fully loaded Turkish Airlines DC-10 took off from Orly Airport in Paris. At 12,000 feet the craft experienced a catastrophic decompression and 346 people lost their lives. Most of the wreckage was strewn over a wide area and none of the bodies was identified, except for six victims who were all found near the rear cargo door, 6 miles from the rest of the wreckage. The Turkish baggage handlers remembered having difficulty shutting the door and did not look through the sight hole. They did not know to look since the directions were written in English which they could not read.

Such disasters always demand a sacrifice, and the investigation focused on Dan Applegate. This is without question unfair. Applegate was, by all accounts, an excellent engineer, but he experienced a "career-ending move" by choosing to do nothing once management had ignored his memorandum. His decision deserves scrutiny because it was his "career-ending move" [10].

Applegate's memorandum came to light 2 years after the crash of the Turkish airliner. By that time Dan Applegate had retired from Convair and became a recluse, protected by his relatives against an outside world that wanted to talk to him about his decision. There is no personal information available on Applegate, and there are no photographs.

Chalkley Hatton (1860–1951)

Fig. 3.17 Chalkley Hatton

What responsibility do engineers have to the public when they have no *legal* responsibility? Actually, the answer is not that difficult. Any profession, be it law, medicine, or engineering, empowers an individual with special talents that benefit the public, and the wise use of these talents for the public good is expected. To do otherwise is to be professionally immoral. But there is a hazy line between moral responsibility and legal responsibility.

This was the question confronting T. Chalkley Hatton, who was one of the early sanitary engineers, along with the likes of Greeley, Hanson, Eddy, Mohlman, and Hazen. Hatton's career, however, had a severe bump in the road, caused by the failure of the Austin Dam, a fascinating story of greed and mismanagement.

The story begins with the construction of a paper mill on the banks of Freeman Run, a small creek in the Susquehanna watershed in north central Pennsylvania. The owner of the mill, one George C. Bayless, kept expanding the mill. By the year 1910 over 200 local workers from Austin were employed at the plant. Further expansion of the mill was restricted by the uneven supply of water in the small creek, and Bayless decided to build a dam that would provide a sustained supply of water for his pulping operations. He hired T. Chalkley Hatton, a civil engineer based in Wilmington, DE to design the dam. Hatton's dam was to be a gravity-type concrete structure, 544 feet long and with an expected rise of 45 feet above the creek elevation. The reservoir was to hold 200 million gallons of water.

From the beginning, the relationship between the owner, Bayless, and the engineer, Hatton, was a rocky one. While Hatton wanted to build a dam according to prevailing engineering standards, all Bayless wanted was to save money. He often overrode design specifications and even ignored Hatton by having the site engineer change the design parameters without telling the home office. For example, Bayless got the engineer to eliminate a valve for a 36-in. drain pipe that would allow for the release of water in an emergency. Instead of installing a valve on the downstream side of the dam, the drain pipe was blocked by an underwater cap on the upstream side. Hatton explained to Bayless that this would make the use of the emergency line quite difficult, but Bayless refused to spend money for the valve. In another instance, the original design had a cutoff wall, designed to prevent seepage under the dam and thereby preventing the dam from sliding on its foundation, but the owner argued that because the dam was built on solid rock, the cutoff wall was not necessary. Finally, the owner, without consulting Hatton, changed the final elevation of the spillway by adding 2 feet of freeboard.

The exasperated engineer finally wrote to Bayless on the first of November 1909:

> Last night I received a telegram from Mr. Rommel, [the site engineer], stating you desired to raise the spillway for the dam two feet and asking for instructions today. I have made a computation of the structure, based upon increasing the height of the water two feet, and I find that it would be dangerous to the stability of the structure to increase the height of the water above what we have provided, and I send you a little sketch, showing wherein it would be dangerous … I therefore cannot recommend to you any increase in the height of the water above what has already been provided, and cannot make any changes to the dam, unless you instruct me to do so over your written signature, thus relieving me of all responsibility.

The dam, in its clearly unsafe state, was completed in 1909. Two months later, a sudden thaw caused the water to flow over the spillway, and during an inspection of the site, Hatton observed several large cracks in the face of the dam. Two sections of the concrete dam had actually slid about 30 in., creating a bulge in the dam. Hatton immediately notified the townspeople, and went about trying to relieve the pressure on the dam. But the 36-in. emergency line could not be opened because there was no valve, so they dropped a charge of dynamite into the water at the general location of the pipe and blew the cap off the pipe. What this did to the subsequent integrity of the dam itself is unknown.

The immediate catastrophe averted, the objective now was to fix the dam. Hatton contacted Edward Wegmann of the New York Aqueduct Commission and the two of them wrote a study in which they agreed that the dam was unsafe and needed to be strengthened. Specifically, they recommended piling stone and rubble against the downstream side of the dam, thus reducing the chances of failure by slippage, and installing a proper cutoff wall to bedrock. Having written the report, Hatton washed his hands of the whole project, and Bayless, the owner, ignored the recommendations.

During the summer of 1911 water was again allowed to collect behind the structure, and by 30 September 1911 the water had reached spillway height. On that day, the lumber mill, located between the dam and the town, was working at capacity and the people of Austin were going about their business unaware of the

Fig. 3.18 The Austin Dam during the winter of 1910/11. Note the severe bow in the dam alignment. The two guys up front are unidentified, but one of them could well be T. Chalkley Hatton. [Photo courtesy of the Potter County Historical Society. Used with permission]

impending disaster. Suddenly at 2:15 in the afternoon the factory whistle blew, warning the town that there was a problem. Some people ran for higher ground, while many ignored the whistle as another false alarm. Shortly after the whistle, a huge wave of water and logs from the mill surged through the town, destroying homes and businesses, leaving the town utterly destroyed. On that sunny afternoon, 78 people lost their lives and several thousand lost everything else.

After the disaster, the magazine *Engineering News* sent an investigator to the site, and his report, based on the location of various sections of the broken dam, clearly showed how the dam had slid off the foundation. The technical reasons for the dam failure were easy to determine, but the human reasons were more complicated. To his credit, T. Chalkley Hatton took the blame for the disaster. He wrote in *Engineering News*:

> The failure of this dam was not the result of poor workmanship, but poor judgment upon my part in determining its foundation. I should have sought the advice of a man more skilled in determining foundations for dams than myself … The great mistake I made in building this dam was trusting the rock foundation to be impervious [11].

In the article Hatton did not blame George Bayless, the owner, and recognized that the fault was his for not insisting that the dam be safe. He apparently was absolved of legal responsibility, and there is no record of his being sued for dam-

(a)

(b)

Fig. 3.19 Austin, PA, (**a**) before and (**b**) after the disaster. [Photos courtesy of the Potter County Historical Society. Used with permission]

ages or being booted out of the ASCE. In fact, Hatton went on to become a respected sanitary engineer, including serving a term as the president of the American Society for Municipal Improvement (the forerunner of the American Public Works Association) and becoming the chief engineer of the sewage commission for the city of Milwaukee. In that capacity, he also helped to write a dictionary of terms used in sewage treatment [12].

Today we would conclude that Hatton did not act properly and that he abrogated his professional responsibility by "washing his hands of the entire project." We now know that an engineer cannot walk away from professional responsibility when the engineer knows that doing so can harm the public. There is, for all of us, regardless of our circumstance or position, the admonition to hold paramount the health, safety, and welfare of the public [13].

3.2 The Morality of Civilian Engineering

Engineering in the service of civil society seems to have little if anything to do with morality. And yet, almost all engineers working in the civilian sector have moral decisions to make on a daily basis. This is not obvious, because when we discuss engineering ethics, we spend time reading about well-known cases such as the DC-10 or the Challenger. Because of this, we might assume that moral decisions in engineering are rare simply because such spectacular events are rare. The fact is that most stories of ethical decision making in engineering are seldom made public. To illustrate, below are three short stories of engineers I have known personally who have had to make ethical decisions. I choose these three simply to illustrate that engineers are confronted with ethical dilemmas almost every day and these decisions are neither publicized nor documented.

- Some years ago an engineering friend was working for a large consulting firm but dreamed of having his own company. One of his jobs happened to be at a location far removed from the main office so that he had to travel there frequently and stay for weeks at a time working with the client. Having some time on his hands, he started talking to potential clients, but instead of bringing the work to his firm, he began to do the work on his own without notifying his bosses at the main office. Soon he had a thriving little business going with his own firm, while he was also drawing a full salary from his other firm. By chance, his bosses discovered what he had been doing and fired him on the spot. He was a talented engineer, and he had accumulated enough work to make it through a rough year, so he landed on his feet. The tragedy was that he never understood, or never wanted to understand, that what he had done was highly unethical.
- A second engineer was a friend who told me about a job he almost had. A developer was converting a large apartment building into luxury apartments and wanted this engineer to do the structu and utility drawings for the conversion.

The engineer asked what was to become of the low- and middle-income families who were then living in the apartment building. The developer had no idea, nor did he care. It was not his problem. He had bought the building, it was now his, and he wanted to convert it into luxury condominiums. My friend the engineer told me that he thought about it for a long time and finally decided that he could not accept the job and be party to the displacement of these people from their homes. There were of course many engineers who had no such concerns, and the building has since been converted into condominiums. There is no record of what happened to the people who were living in the building.

- The third engineer is also in private practice. Some years ago I asked him to come to my engineering design class and talk about what it is like being a real engineer. The conversation went well, and, encouraged by the interest and responsiveness of the students, his presentation got increasingly personal. Finally he said that there was one incident that he had not told anyone about and that he wanted to share with us. It seems that his firm had been working with a local developer who had greatly overextended himself and was in deep financial trouble. One day the developer asked the engineer to meet him at the bank for a conference. It turned out that the developer had found some local people, known to both the developer and the engineer, who were willing to invest \$250,000 in the development company. As they were transferring the money to the developer the engineer knew that the developer's financial situation was grave and that he owed millions of dollars and was without doubt going into bankruptcy. A mere \$250,000 was a drop in the bucket and would not have made any difference in the eventual success or failure of the development corporation. The investors were almost certainly going to lose their money. The code of ethics, however, required the engineer to keep quiet. But he had not forgotten the incident and he worried about it. After he had finished his story, he asked our engineering students, "What should I have done?"

In all three of these cases the engineers' decisions were out of their technical fields. Was the first engineer ever taught about loyalty to an employer while he was in engineering school? If the engineer who refused to work on the apartment house conversion had thought that the "public" was the displaced families, then he certainly had a moral right to refuse the work. But were not the owner of the building and the people who would buy the condominiums also the "public"? Did he have any moral justification for not accepting the job? Finally, the third engineer could not prevent the loss of the life savings of his friends because of his commitment to be a faithful representative of his client. What about his personal moral concerns? Should those override his professional responsibilities? How could he have justified his actions to the local investors after they had lost their money?

None of these stories will ever make the local newspaper, much less become case studies for us to ponder, and yet these engineers, all engaged in civilian engineering, made ethical decisions that affected their clients, the public, and, no doubt, themselves.

References

[1] Kidder R (2005) Moral courage. Morrow, New York
[2] Bowman G (1966) John Smeaton – consulting engineer. In: Engineering heritage, vol 2. Dover, New York
[3] Yin K-H (2007) Outstanding women in mechanical engineering, *Int J Mech Eng Educat* 35(3)
[4] Graham L (1996) The ghost of the executed engineer. Harvard University Press, Cambridge, MA
[5] Morgenstern J (1995) The fifty-nine story crisis. *The New Yorker*, 29 May 1995
[6] Boyd TA (2002) Charles F. Kettering: a biography. Beard Books, Frederick, MD
[7] "The Amazing Mr. Midgley" *The Nation*, 20 March 2000
[8] Cohen R (1974) A heartbeat away: the investigation and resignation of Vice President Agnew. National Archives and Records Administration, Washington, DC
[9] Eddy P, Potter E, Page B (1976) Destination disaster: from the trimotor to the DC-10. Times Books/Random House, New York
[10] Fielder JH (1992) Floors, doors, latches, and locks. In: Fielder JH, Birsch D (eds) The DC-10 case: a study in applied ethics, technology, and society. SUNY Press, Albany
[11] Austin Dam and its failure. Eng News 68(14):605–607, 1912
[12] Skinner JF, Gascoigne GB, Gregory JH, Hatton TC, Mohlman FW, Stevenson WL (1925) Definitions for sewage and sewage disposal practice. Am J Public Health 15:327–334
[13] Rich T (2006) Lessons in social responsibility from the Austin Dam failure. Int J Eng Educat 22(6):1287–1296

Chapter 4
The Engineer's Commitment to Society

The engineers introduced in Chapters 2 and 3 all had varying degrees of commitment to society. Some took it seriously, like Peter Palchinski, Vauban, and Lucius Clay, while others, like Kurt Prüfer, Abdul Khan, and Lester Matz, did not. The reasons for this commitment, and its evolution, says a lot about the engineering profession.

One of the most interesting engineers discussed earlier is Dan Applegate of the DC-10 story. While we do not know Mr. Applegate very well, we can stereotype him as an example of the technically excellent engineer thrust into a decision-making role that tests his personal and professional ethics. To some degree, Applegate's inaction after his memorandum was rejected by Convair management could have been supported by his own code of ethics, if indeed he even knew there was such a thing. A fundamental canon of the ASCE Code of Ethics states:

> Engineers shall act in professional matters for each employer or client as faithful agents or trustees.

Dan Applegate was faithful to his employer. Loyalty is morally defensible if the object of the loyalty is engaging in moral actions. The Nuremberg trials after the Second World War established that blind loyalty is not acceptable and that "following orders" is not honorable behavior if the action leads to immoral results. But Dan Applegate no doubt firmly believed that his job and his loyalty to Convair were both honorable and moral.

Another part of the engineering code of ethics states:

> Engineers shall advise their employers or clients when, as a result of their studies, they believe a project will not be successful.

Dan Applegate did exactly that. In his mind he no doubt felt that he had done his duty and from now on the responsibility rested with management.

Importantly, however, there is a potentially conflicting statement in the code of ethics:

> Engineers shall hold paramount the safety, health, and welfare in the performance of their professional duties.

If Applegate had considered this obligation, he might have decided to override the first two ethical requirements. At the very least, he had an ethical decision to make [1]. On the one hand he would want to be loyal to his employer, but on the other hand he would realize that when his employer did not resolve the problem to his satisfaction and that his concerns could result in the loss of life, he had the responsibility to go outside his company to prevent the catastrophe he knew was sure to occur. This action is known as *whistleblowing.*

Ethicists writing on whistleblowing make a distinction between *permissive* whistleblowing and *obligatory* whistleblowing [2]. The first test in any situation like that faced by Dan Applegate is whether or not he was permitted to air his concerns outside the company. Usually this permission is valid if the potential harm is significant and if the whistleblower has exhausted all avenues within the organization. The risk of harm in this case was great, and by sending the memorandum to top management Applegate had exhausted his options within the company. He thus had "permission" to take the next step – to go outside the company.

Was he obligated to do so? Most writers argue that the "obligation" test is based on two conditions:

1. The action will not result in great harm to oneself.
2. The action has a reasonable chance of being successful.

If the effect of the action will result in great harm to the whistleblower or if the action has little chance of succeeding, then the action is ethically unwarranted and there is no obligation to blow the whistle. As a most extreme example, suppose a German railroad employee during the Second World War realizes that he is switching trains full of people who will be killed at a concentration camp [3]. Clearly there is no chance whatever of success if he chooses to be a whistleblower, and there is a high probability of his own death.

In the case of Dan Applegate, if he had gone public with his concerns, he would certainly have lost his job. While this is not trivial, it pales in comparison to the loss of 300 lives. A skillful engineer can get other jobs and his fame might even have resulted in enhanced income. In our litigious society, going public with his concerns would have had an immediate effect in making McDonnell Douglas redesign the rear cargo door. Therefore Applegate was, by most ethical tests, obligated to tell someone outside the company in order to get the door and floor redesigned.

The possibility exists that a disclosure would have resulted in such adverse publicity that McDonnell Douglas would have abandoned the DC-10 altogether and thousands of people would have lost their jobs. The obligation is thus not so clear-cut, and it is not easy to blame Dan Applegate. Engineering ethics is a murky, difficult topic, complicated by the fact that these ethical responsibilities are superimposed on top of everyday ethics, the moral responsibilities we hold true because of our common commitment to our society.

4.1 Everyday Ethics

Ethics asks the question: How, all things considered, ought we to treat each other? Thousands of philosophers, working over 25 centuries of human civilization, have failed to satisfactorily answer this question. But we need to try, and we can start by understanding what we mean by morality and moral values, where these values come from, and how they can be put to good use in everyday life. Knowing what moral values and ethics are will not guarantee that good decisions will be made, but it will allow us to recognize that some problems have moral dimensions. Once that is understood, knowledge about ethics can help us to think through such problems.

4.1.1 Moral Values

Morality is a set of rules that ought to govern how people treat each other. These rules are accepted by rational people because they recognize that doing so will be to their mutual benefit. While we might disagree on details, it is clear that we will agree on the most important moral values, and that these same values appear in almost all cultures and traditions. Take for example the morality of telling the truth (or, in the negative, of lying). Telling the truth was among the Ten Commandments that Moses presented to the Israelites: "Thou shalt not bear false witness."

In a more modern version of the Ten Commandments Dartmouth philosopher Bernard Gert suggests that moral values that all people share can be summarized as follows [4]:

1. Not killing others
2. Not causing pain
3. Not disabling
4. Not depriving freedom
5. Not depriving pleasure
6. Being truthful
7. Keeping promises
8. Being honest
9. Obeying the law
10. Doing your duty

Most people will agree that not being killed is good, that having freedom is good, and that experiencing pain is bad. Truthfulness and kept promises are good, as is honesty (not cheating), obeying the law, and doing one's duty. One may quibble about the details of these values, and we might want to suggest others that we personally find more fitting to be on the list, but all in all, this is not a bad start. Moral rules such as the list above are the result of rational thought and consideration of how we best might get along. It is clear that stealing from each other is not a good idea, and that we are all better off if everyone respects everyone else's property.

The idea that it would be to everyone's benefit if moral rules were universally accepted is called a "social contract" and was first proposed by 17th-century British philosopher Thomas Hobbes. He argued that this contract is what makes civilization possible, and that in the absence of such an agreement, we would experience continual fear and ever-present danger of violent death, and that our lives would be "solitary, poor, nasty, brutish, and short" [5]. An immoral person, then, is one who does not abide by the rules of the social contract, and if everyone acted in this way we soon would descend into Hobbes' "state of nature" and civilization would cease.

But we can argue in response that none of us ever signed the "social contract," such as agreeing to return lost property to the rightful owner. Instead, such actions are *implicit* in a well-functioning society. We would all benefit if everyone acted in this way, and we would certainly lose the benefits of civilization if people did not act morally.

Doing the right thing in everyday life should be automatic. A good analogy is the use of grammar in a language. None of us has signed an agreement to use English grammar in a certain way, but we do, and such use is automatic. We use correct grammar because we want people to understand us. If we talked gibberish, there would be little reason to have any social interactions and we would lose the easiest method of communicating with each other. Grammar comes automatically to us. We don't have to think ahead to construct a sentence: "need a noun, and verb, and the adjective has to modify the noun," and so on. We just know what we want to say, and out it comes. Living a moral life is very similar to the use of grammar. Once you learn how to make the right decisions, you begin to act that way automatically.

One common moral problem is telling the truth. Everyone has lied at one time or another, and although we have recognized this as less than honorable behavior, we usually manage to justify it, at least to ourselves. What makes telling the truth (or in the negative, lying) a moral problem? Consider a typical situation. An engineer has padded her resume so as to appear more experienced than she really is in the hopes of securing a contract to do some design work. She has not told the truth. Why is this a moral problem? She can argue that she feels perfectly competent to carry out the design, and that nobody will be hurt by the padded resume.

What would happen if everyone did this? That is, if this practice were universalized? If it were no longer possible to believe any resumes, the system of choosing the most qualified engineer would break down, and this would without doubt harm individuals in society. Because it would be harmful, resume padding is thus immoral. We have to conclude that lying is morally unacceptable. For an engineer to lie is not just a personal decision, but one that affects other people.

4.1.2 *Promoting and Supporting Moral Values*

The reason we have to promote and support both personal and professional ethical behavior is because we can predict what would happen if people did not behave morally. Yet some will argue that there are no such shared moral values, and that every society creates its own set of values, making no single list of moral behavior applicable to all.

Cultural relativism is an easy trap to fall into. It is tempting to argue that morals are nothing but cultural traditions, and in some cultures these values are different. This theory maintains that there are no universal moral values, and every culture's values are just as good as those in other cultures.

If we accept the proposition that morality is culturally relative, however, we have to accept some troubling conclusions. First, if the morals of any culture are just as good as the morals of other cultures, then there are no "good" morals or "bad" morals. That is, each value is just as good as any other as long as it is accepted by that culture.

People who support cultural relativism hold that values are relative. This must be true because different moral values exist in different cultures. For example, they point to the treatment of women in Saudi Arabia (who are not allowed to drive a car, vote, or go to school) and conclude that this proves the relativity of morals. Such discrimination would be highly immoral (and illegal) in most Western countries, but it seems to be widely accepted in Saudi Arabia.

But do such differences prove that moral values are culturally relative? Suppose killing each other with impunity is a part of a culture. That is, there is no "Thou shalt not kill" in that culture. Is this a good situation, and would you personally want to live in such an environment? If you would not, what makes you think others would? Would you believe that people in some cultures would be quite content to accept uncontrolled killing? Yes, of course, we have cultures (Somalia comes to mind, for example) where people constantly fear for their lives, but is this a good way to be?

A second problem with cultural relativism in morality is that if we accept this as the truth, we cannot condemn others for doing what we would believe to be immoral. We could not, for example, condemn the segregationists during the Civil Rights era as holding immoral values. But we want to say that segregation and racial discrimination are *wrong* regardless of who believes they are right. If we accept cultural relativism as true, we cannot say that some moral values are better than others. Belief in anything, such as segregation, does not make it right, and therefore some cultural values must be morally wrong.

A third problem with accepting cultural relativism in moral values is that if it is true, then there can be no moral progress. Any moral value is then perfectly good for that time and place, and there is no need for improving the moral environment. We would not, for example, have any reason to criticize the treatment of women as property. But we all agree that such treatment of women is patently immoral, and we are pleased that we have, in this country at least, come a long way toward fixing this wrong.

The *value* of the concept of cultural relativism is that it makes us question our own moral values. Suppose the government of the USA decided to emulate more restrictive societies by censoring books and newspapers, arguing that this is necessary for protection against terrorism. Such censorship would break several of Gert's moral rules (e.g., do not deprive freedom, be truthful, among others), and we would have to ask if this is acceptable behavior for our country. We would conclude (or at least we have concluded up till now) that censorship is too severe a price to pay for the incremental safety that such a policy would provide.

So we are left with the conclusion that basic human moral values are shared, or at least that all humans would like to share them. Some people are unfortunate to live in societies where these values are not respected, but almost all would, if they could, choose to live in a society that shared such values. During the Cold War years many people risked death and imprisonment trying to escape to the West. The East German Communists, not the West Germans, built the Berlin Wall. The barbed wire on top of the fences was designed to keep people from fleeing from an immoral society to one that respected moral values.

4.1.3 When Moral Values Conflict

Morality would be simple if in all cases a single moral value were in play at any give time. But this is seldom the case. It might not be possible, for example, to tell the truth and keep a promise at the same time, and so we need a way to analyze situations in order to determine what, all things considered, is the right thing to do. The search for a method or model that yields the right answer is called *normative ethics*. If telling the truth would result in breaking a promise, or if being loyal would result in the death of innocent people, then there is a moral dilemma. One purpose of ethical models is to help resolve such dilemmas. The distinction between ethics and morals becomes quite clear.

> *Morals* are those values that we adopt in order to best get along with each other, and *ethics* is the means of resolving conflicts between moral values.

Ethical theories are models constructed to help resolve conflicts between competing moral values. There are thousands of such theories, but the most widely respected ethical models fall into two groups: *consequence-based* and *duty-based* ethics [6].

Consequences. In the first group of normative ethical theories, the rightness or wrongness of an act is judged on its consequences: how much good or harm would result from an act. One form of a consequentialist ethics is called *hedonism*, the principle that one ought always to act in such a way as to maximize the benefits to oneself. People who act in this way soon recognize that short-term hedonistic actions that may seem at first to bring one happiness are often the cause of much unhappiness later. Hedonism as a tool for making ethical decisions is a blunt object at best.

Probably the most famous consequentialist ethical theory is *utilitarianism*, in which the best act is one that results in the greatest sum total of happiness (or pleasure) to everyone involved. An act is ethically right if it leads to the greatest total happiness, where the happiness of everyone involved in the decision is calculated.

Engineers are instinctively drawn to utilitarianism because it seems to offer an objective and reproducible method for making ethical decisions. The concept of benefit/cost (B/C) analysis for making decisions is a utilitarian idea. If it is necessary to decide where to put a dam, for example, the cost of the dam at the various locations is calculated and the benefits that would be attained from the dam locations are estimated. The benefit (B) is then divided by the cost (C) to obtain the B/C ratio. The dam location with the highest B/C ratio should be the clear winner. It is tempting to think that a similar calculation is also possible for ethical problems so that the right answer can be calculated from several alternative courses of action.

While utilitarianism is an attractive ethical concept and often yields suggested actions with which we can feel comfortable, it also has pitfalls, one of which is that we cannot know for sure what will happen in the future. We may do what we feel is a wonderfully inventive and thorough calculation, only to be blindsided by events that we could not have predicted.

A second problem with utilitarian ethics is that if we try to maximize the happiness (or pleasure or fulfillment or some such parameter) for everyone involved, clearly wrongful acts can appear to be right. Consider the euthanasia program instituted in Nazi Germany where mentally retarded and other handicapped people were killed in order to enhance the gene pool. In the utilitarian calculus, a quick death is of no moral concern as long as there is no physical or mental pain. Once a person is dead, he or she is no longer counted in the summation of happiness/pleasure, and thus killing the infirm results in a positive happiness number. But we cannot agree with this result; we would argue that intentionally killing innocent humans is an evil practice, no matter what the utilitarian calculus might suggest.

On the positive side, utilitarianism is one of the few ethical systems that can accommodate creatures other than humans in its decision making. If pleasure is the yardstick to use as the positive outcome, then pain is the inverse (just as in B/C calculations), and we can use the pleasure/pain ratio to make ethical decisions. For example, we can argue that using spring-loaded leg traps for trapping animals such as fox, coyote, beaver, and other creatures is a cruel and evil practice only if we consider the pain suffered by the animals. If animal pain is not considered, then the use of leg traps would not be immoral because only the trapper sees the animal, and he of course gets pleasure from trapping. We might argue that the pain suffered by an animal ought not to be considered equal to the pain suffered by a human, but this is quibbling. Our argument would be that pain is pain, and causing such pain is an evil act.

Duties. An alternative approach to ethics is to argue that it is not consequences that matter, but it is the act itself that is important. This is called *deontological* or *duty-based* ethics, and the objective is to do that which is considered right, regardless of consequences. For example, we could argue that telling the truth is such an important moral value that we should always tell the truth regardless of the consequences, that we have a *duty* to tell the truth at all times.

The Ten Commandments is an early version of duty ethics. In this case one has a duty to behave in a certain way because God commands it. But we don't have to base our duties on God's commandments (if in fact we knew exactly what God wanted us to do).

A famous and widely admired ethical system using duties was developed by Immanuel Kant, an 18th-century philosopher. He argued that one has a duty to behave in such a manner that one can support a rule that should be applied to all people in similar situations. For example, if an engineer is tempted to take a bribe, he must ask if it would make sense to have a rule that all engineers, in a similar circumstance, ought to have the right to take bribes. That is, is this engineer sufficiently different in some significant way that we could make a rule that would allow him to take bribes but would prohibit others from doing so? Kant called this idea *universalizability*, a cumbersome word that represents that idea of being able to make a rule that can be universally applied. This idea often helps resolve personal conflicts in deciding whether some action is or is not moral.

Another way of deciding on the morality of actions is to recognize that in some situations moral rules conflict, thus leading to the breaking of one or more moral rules. The moral rules suggested by Gert are useful if only one rule defines the decision (e.g., keep promises). But suppose it is impossible to solve an ethical problem without breaking a moral rule. For example, you promise to meet a friend for lunch, but there is an emergency and you have to drive another friend to the hospital. You cannot both keep the promise and do your duty. Gert's solution to such a dilemma is to suggest that if any of the moral values listed above conflict, the choice should be based on what an impartial and rational person would do in a similar situation. A rational and impartial person would agree that taking a friend to the hospital is more important, and breaking the moral rule to keep promises is the right thing to do.

Such an ethical scheme invites the question of what exactly is an impartial and rational person, and this is open to argument. On a practical level, using the technique of another person to guide one's decisions is a good one if the rational, impartial person has a reputation for making good decisions. An engineer who is a moral exemplar might be a good model, and when confronted with an ethical decision we might ask what this person would do in a similar case. We may never know for a fact what he or she would do, but wanting to emulate a person of worth whose opinions one respects is not a bad technique for making ethical decisions.

Virtue. Ethical models such as consequentialist and duty-based models are all intellectually interesting, but they often do not yield answers with which we are comfortable. If classical ethical models provide unsatisfactory answers, then maybe we have been asking the wrong question. Maybe the question is not "what, all things considered, should I now do?" but rather "how should I live?" One of the first and greatest philosophers of all time, Aristotle, took that approach, and many people today believe that the question of "how should I live?" is still a useful way to decide moral dilemmas.

Aristotle believed that one ought to live a virtuous life. A person of good character, or a virtuous person, not only does what is right in various circumstances, but also is a good citizen in his or her community. This inner integrity is what drives a person to do the right thing without even thinking about it, and the virtues that a person has are properly balanced so that good decisions emerge. For example, the virtue of courage is nicely balanced with cowardice, in that a reasonable person would not want to be too courageous (trying to stop a speeding train by standing in its way) nor would he or she want to be too cowardly (being afraid to speak out in support of moral behavior). A good balance, or "the golden mean," is what a virtuous person desires.

If we list the virtues that Aristotle thought worthy of a good citizen, we come to essentially the same list of moral values we started with – Gert's "Ten Commandments" –completing a loop. Holding to these values is the same as being a virtuous person, and the world would be a better place if all people behaved virtuously.

4.2 From Personal to Professional Ethics

In 1914, some members of the American Society for Civil Engineers wanted to adopt a professional code of ethics but were talked out of it by men who argued that personal ethics were sufficient to govern professional conduct. It took another 50 years for the society to realize that professional ethics is another layer of ethical responsibility on top of personal ethics. While the basic ideas of professional ethics are similar to those of personal ethics, the difference is that professional ethics come into play because of the engineer's commitment to society.

Engineers are technology experts, unlike most people who have no idea how the technology they use every day actually works. Most people drive their cars, turn on their televisions, use their computers, and withdraw money from automatic teller machines, but they have no idea how these systems work. They rely completely on engineers to ensure that these systems work. Of course, no engineer can be an expert on everything, so engineers themselves rely on other engineers. The difference is that technology for an engineer is never mysterious and any engineer can quickly explain a highly technical problem in his or her own specialty to any other engineer.

Engineers, as other professionals, are necessary to the functioning of society. If knowledge is to be made useful, it is the engineer who must do it, and engineers are absolutely essential to getting things done. For example, take the cleanup of a hazardous waste spill. A team of hydrologists, biologists, geologists, anthropologists, and other specialists can analyze the situation and decide on a course of action, but in the end they turn to the engineer to actually get it done.

Because of this societal responsibility, engineers have certain obligations in performing their duties. Most importantly, they have an obligation to do the right thing, even if it means losing a job or incurring some other cost. The engineer's role carries with it responsibilities exceeding those of a private citizen. Not only are engineers expected to act morally as people; they must act morally *in their role as engineers*. Therefore we have *professional engineering ethics,* an additional layer of responsibility on top of personal moral responsibilities.

The moral rules discussed above have a parallel in engineering. These rules are called *codes of ethics*, and the code of the National Society of Professional Engineers is typical of engineering codes of ethics. The first section, or canons, of the NSPE Code of Ethics is as follows:

Engineers, in the fulfillment of their professional duties, shall:

1. hold paramount the safety, health, and welfare of the public,
2. perform services only in areas of their competence,
3. issue public statements only in an objective and truthful manner,
4. act for each employer or client as faithful agents or trustees,
5. avoid deceptive acts, and
6. conduct themselves honorably, responsibly, ethically, and lawfully so as to enhance the honor, reputation, and usefulness of the profession.

The full code goes on to add rules of conduct that refer to these fundamental canons.

Just as the ten moral rules for everyday life listed above is a good first step at how to live a moral life, so a list of fundamental canons is a useful set of principles for making morally laden decisions in engineering. For example, we would agree that engineers ought to practice only in their areas of competence, and we would agree that engineers should not give or take bribes. So while professional engineers might disagree on the details of the code of ethics, most would argue that such a set of fundamental canons is good and useful.

The engineering code of ethics, although it might vary from discipline to discipline, is basically the same document and attests to the fact that engineering is a profession. The story of its development is useful for understanding the engineer's ethical responsibility to society.

4.3 Engineering Codes of Ethics

Each engineering subdiscipline has its own code of ethics. Strictly speaking, these documents are not codes of ethics, because ethics is the process of careful deliberation of the right and wrong thing to do in a given circumstance, and this cannot be reduced to a code. The engineering codes of ethics are actually lists of guidelines spelling out to the practitioners and the public the responsibilities of the individual professional. As such, a better title for the codes of ethics would be *guides to responsible conduct*, but since they are popularly know as codes of ethics, we will continue using the designation in this discussion.

Engineering as a profession saw a stepwise development in its code of ethics. The earliest codes addressed ethical concerns between and among fellow engineers and included such rules of conduct as "do not steal another engineer's client" and "do not speak disparagingly about a fellow engineer." With time, the codes incorporated the engineers' duties to employers and to clients, and such rules as "be loyal to your employer" were added. Then in the 1970s the recognition that engineers held a moral responsibility to the public led to the addition of the famous clause "engineers shall hold paramount the safety, health, and welfare of the public."

Engineering codes of ethics have two types of statements that reflect morality: *admonitions* and *requirements*. Admonitions are statements that strive to lead the engineer to the moral high ground, to make the engineer design his or her professional life so as to routinely act with moral integrity. Admonitions are statements specifying what the engineer ought to do to be a good engineer. To *not* adhere to an admonition statement in a code of ethics will not get engineers in trouble, but to adhere to the admonition will make them better engineers. For example, a code might include the following:

> Engineers shall continue their professional development throughout their careers and should keep current in the specialty field by engaging in professional practice, participating in continuing education courses, reading in the technical literature, and attending professional meetings and seminars.

Nothing bad will (probably) happen if the engineer ceases to learn, but a better engineer is one who stays current. The reason this is a *moral* statement is that engineering involves public welfare, and not keeping up with technical developments could result in an incompetent design that may harm the public.

The code of ethics is also full of requirements, things the engineer *must* do to continue to be part of the engineering community. The difference between admonitions and requirements is that not following the requirements *can* result in harm to both the engineer and the public. Ignoring the requirements would be to act immorally in professional engineering.

For example, a code might include the following:

> Engineers shall not affix their signatures to any plans or documents dealing with subject matter in which they lack competence, nor to any plan or document not prepared under their direction and control.

Suppose a civil engineer who knows little about electrical circuits approves drawings for the wiring of an elevator for a building. If the elevator fails and people get hurt, the civil engineer could not plead ignorance of electrical circuits, and this engineer would be legally (and morally) at fault.

4.4 Limitations of Engineering Codes of Ethics

The codes of the various engineering professions are quite similar, and all provide a first line of defense when ethical questions arise. With enough diligence and information, most ethical problems can be solved using the basic premises of the codes. But sometimes the codes are unclear about details of circumstances, or they even give contradictory answers.

One problem with the codes is that the "public" to whom the engineers owe primary responsibility is not defined. This problem is especially acute for engineers working in the armaments industry. Their primary mission is to develop technology to kill the most "bad" people while protecting the "good" people. But if people are people, then are they all not part of the "public" to whom the engineer owes responsibility?

Sometimes engineers in economically wealthy countries with strict public safety rules and regulations are tempted to ignore such regulations when conducting business overseas. The manufacture and sale of products banned in the USA but legal elsewhere can cause serious ethical problems. For example, sale of a banned pesticide requires a definition of just what "public" the engineer is responsible to. If the public is the people in the developing country, short of food and needing inexpensive pesticides, then the engineer is well within his moral rights. If the public includes children who will be harmed by ingesting the pesticide, then the right thing to do is less clear. And if the public is the people of the world who will be negatively impacted by the use of a persistent pesticide wreaking havoc with natural ecosystems, then the action would appear to be immoral. Note that nothing in this discussion thus far is about whether or not an action is legal or illegal. More on that later.

The point is that no set of ethical rules can anticipate every situation, and many questions in engineering ethics require considerable thought. It is foolhardy to believe that every question will be answered by looking it up in the code of ethics.

Finally, the engineering codes of ethics have little to say about questions regarding the environment. No code spells out what responsibility, if any, engineers have to nonhuman animals, plants, or places. The only concern is that the engineer's actions not diminish the welfare of the (human) public. If an engineering project causes the demise of an animal or plant species, the concern is not for that plant or animal, but for future humans who may not be able to enjoy looking at this species or obtain some beneficial use from it.

The conclusion we come to is that the engineering code of ethics is a fine first, and very rough, tool for making ethical decisions in engineering. Often when engineers are confronted by ethical problems, a quick glance at an engineering code of ethics is enough to encourage a decision that the engineer can live with. But ethical problems are seldom straightforward, and right actions are not obvious. There is a great deal of subtlety in ethics, and any set of guidelines such as a code of ethics cannot hope to cover all cases. As an example, let us consider the first, and arguably the most important, canon of almost all engineering codes of ethics.

4.5 The First Canon Reconsidered

The first canon of many engineering codes reads:

> The engineer shall hold paramount the safety, health, and welfare of the public.

This seems iron-clad, with the two words "paramount" and "shall" making this mandatory. But are there ever reasons for violating this canon in the performance of engineering duties?

There are at least four arguments that might be advanced by someone who has violated the first canon of the code:

- The requirement is internally inconsistent.
- My religious convictions compel me to do otherwise.
- I don't think the public really knows what is in its best interest.
- I was forced to do it.

Let us consider each of these in turn.

The requirement is internally inconsistent. By stating that the engineer shall hold paramount the health *and* safety *and* welfare of the public, the implication is that the statement has internal consistency. That is, it is possible in all cases to hold all three of these requirements paramount simultaneously. There may be situations, however, when the engineer, in order to fulfill the obligation to uphold the safety of the public, will not be able to simultaneously promise to provide for the public welfare (in the opinion of the public). An example might be the 55-mile-per-hour speed limit imposed by traffic engineers (and others). Engineers believed that this lower speed limit was in the public interest, saving lives and reducing gasoline consumption. But the cost in travel time was too much of a cost for the public, which measured its welfare on the basis of how fast they could get from one place to another. Another example might be the conflict engineers faced a few years ago in the construction of apartment houses in the rapidly industrializing regions of Turkey. Their commitment to public welfare – providing places to live – apparently outweighed their concern for public safety, and many houses were apparently built with substandard materials and shoddy construction.

In cases where the three requirements – safety, health, and welfare – might conflict, the engineer is placed in a conundrum. The only morally acceptable way out of such difficulties is to remember that the social contract engineers have with the public requires the public to specify just how much health, how much safety, and how much welfare it wants. In the case of the 55-mph speed limit, the public spoke and engineers listened. In the case of the apartment houses in Turkey, it is unclear if the engineers who designed the apartment houses ever asked the occupants if they wished to live in houses that might collapse and kill them should an earthquake occur. If they had, and if the people had agreed to assume the risk, the engineers would have been morally justified in constructing apartments that under normal conditions functioned perfectly well as living spaces.

My religious convictions compelled me to do otherwise. Of the many religions in the world, a few are used here to illustrate how an engineer's decisions might be influenced by religious beliefs. For the "religions of the book" (Judaism, Christianity, Islam) the word of God can be found in a written text. Unfortunately, God is not very explicit about roles of engineers. While it is quite clear that God does not allow round haircuts (Leviticus 19:27) or the sowing of two kinds of seeds (Leviticus 19:19) or the wearing of garments from two different kinds of material (19:19), it is not possible to find an *unless* clause that could be added to the first canon. Accordingly, the faithful have to rely on interpretations by their religious leaders. This should not be taken as somehow wrong. One could argue that when we have a toothache we go to the dentist, or when we have a legal problem we go to a lawyer. Why, then, if we have a moral problem concerning one's role as an engineer, should we not go to the preacher/pastor/rabbi/mullah/priest/etc.? Are these people not professionals, paid to resolve moral problems? But the *unless* clause that various religious leaders might suggest would be quite different.

If confronted by the question of whether *paramount* has any qualifiers, Judaic, Christian, and Islamic religious leaders might say, yes, that the rule should be:

> The engineer shall hold paramount the safety, health, and welfare of the public, unless God (Allah) says otherwise.

Invariably, people who believe in such religions will know what God intends based on some passage in the sacred book. Once an engineer knows what God wants, then such requirements would modify the *paramount* clause and cause engineers to change their professional actions. Engineers who subscribe to the Roman Catholic Church's stand on birth control, for example, might find it difficult to work on birth control technology.

Other major religions do not depend on the written word but rather rely on long-standing traditions. The Buddhist tradition considers the whole world in its organic sense, with nothing existing in isolation, and with everything connected to everything else. Humans achieve a harmonious relationship with nature by exhibiting proper humility and caring. Buddhism has as its main tenet the principle of *ahimsa*, do not destroy life, and teaches compassion for all of life, including trees, forests, and wildlife.

Hinduism maintains that all animals are incarnations of other living things, including people, and that even the gods were at one time monkeys, cows, or other creatures. This belief leads to a reverence for certain animals and a prohibition against eating meat. Hindus see humans as a part of the total environment, not as separate from nature. A proverb seems to sum up the Hindu religion: "Do not kill any animal for pleasure, seek harmony in nature, and lend a helping hand to all living creatures" [7].

Thus the Buddhist or Hindu engineer might argue that the first canon should read:

> The engineer shall hold paramount the safety, health, and welfare of the public unless such action destroys life.

An example might be an engineer who is asked to design facilities for a big-game ranch in which large animals are brought to a fenced compound only to be shot by "sportsmen." While Christians would not find much in the Bible to prevent such activities, a Hindu engineer would not want to be part of killing for pleasure, even if this is for public welfare ("entertainment").

In Japan, the Zen Buddhist movement also stresses the oneness with nature and the merging of self within one's experience. Nature is around us, and we must contemplate and understand it as well as learn to appreciate its beauty. Such appreciation and self-awareness in the Zen tradition leads to fulfillment. The Zen engineer might then say:

> The engineer shall hold paramount the safety, health, and welfare of the public unless such action detracts from the beauty of the earth.

The construction of Interstate 40 across the pristine Smoky Mountains in North Carolina is an example of a project that detracted from the beauty of the earth. The forest through which the Appalachian Trail meanders is one of the largest and most beautiful in the world. Now, with Interstate 40 crashing through the middle of the forest, it becomes just another place to park an RV. An engineer who appreciates Zen philosophy would mourn the destruction of beauty and would not see this as a public benefit.

Most ancient religions, including Polynesian and Native American religions, are animistic, and recognize the existence of spirits within nature. In such religions the spirits do not take human form, as they do in Greek or Roman religions. Spirits simply are within the tree, the brook, or the sky. It is possible to commune with these spirits, talk to them, feel close to them, and thereby feel close to nature. In many animistic religions the killing of an animal such as a deer or bear requires the proper appeasement of that spirit. Cutting down a tree requires an explanation to that tree (spirit) as to the reason for cutting it down, and the explanation better be a good one or the spirit will haunt the cutter long after the event. It is not too farfetched to say that, if an engineer believes in the animist traditions, the first canon might be changed to read:

> The engineer shall hold paramount the safety, health, and welfare of the public unless such action offends the spirits of the earth.

In Iceland, a substantial number of people say that they believe in the existence of "little people" who live in rocky crags and deep woods. When laying out roadways, engineers respect these beliefs by avoiding places where homes of the "little people" are believed to be. Icelandic engineers act to achieve the greatest good by recognizing the spiritual component in the lives of the people they serve.

In summary, it is likely that if the engineer believes that his or her most important goal is to follow the dictates of religious beliefs, then the first canon would have to be modified and the engineer would consider religious consequences to be paramount.

There are many arguments against such a conclusion. The simplest is to reiterate that ethics concerns the integrity of each person. As long as we respect all humans equally, we must agree that every person is entitled to individual opinions on matters that cannot be proven. That is, one person's religion is just as right as another's, and one god is just as real as any other god. If we truly respect each other's opinions, then we would have to agree that the possibility exists that any one of the many religions invented by humans might be "right," and all others would then be "wrong." If this is a possibility, then it may also be that all religions are "wrong" in the sense that some of the historical traditions may not actually have occurred and that spiritual doctrines might be mistaken. But at the same time all religions are "right" if they provide a sense of personal worth, fulfilment, and sense of belonging.

Thus we cannot know with any certainty what, if anything, God or a religious tradition commands the engineer to do. To supersede the "paramount" clause in the first canon because of a personal religious belief is morally indefensible.

I don't think the public really knows what is in its best interest. The oft-used utilitarian principle of creating the greatest overall happiness can be applied to engineering as well. The maximizing of benefit is an attractive ethical perspective for many engineers, because engineers tend to be utilitarian in their outlook. In the B/C analysis, invented by the US Army Corps of Engineers who in the 1950s had to make decisions concerning competing water resource projects, the dollar costs of each project and the estimated dollar benefits were calculated. For example, they would estimate how many people might want to use a particular swimming area and how much they would be willing to pay for such a privilege. Reduction of losses from floods and the availability of a drinking water supply are two other benefits. If dollars could be equated with happiness, then the projects that had the highest B/C ratio would be the best ones to undertake because they would produce the greatest overall happiness [8].

In the 1970s the Corps of Engineers was genuinely surprised that the public did not always go along with this analysis. The Corps could show how a project had a high B/C ratio, and yet there would be many people opposing the project. The Corps first ascribed this apparently irrational behavior to technical illiteracy. They believed that the public simply did not understand what was important. The Corps told themselves that they were doing the right thing, and that they cared about the welfare of the public. They even started to wear an unofficial button on their uniforms that proclaimed "THE CORPS CARES." Their objective, in the great utilitarian tradition, was to do what they knew to be in the public interest.

If the Corps of Engineers could have altered the first canon, it might have been to something like this:

> The engineer shall hold paramount the safety, health, and welfare of the public even if the pubic does not understand what is in the public interest.

Paternalism is defined as interference with a person with the purported justification that the interference is for the promotion of the person's good or the prevention of harm. Often paternalism involves the loss of liberty, such as helmet laws for motorcyclists, but sometimes it might be as simple as lying, such as when a physician is not truthful with a dying patient. The moral justification of paternalism is that it is done for benevolent purposes. But this is contrary to the moral requirement of individual autonomy – allowing each person to lead his or her own life. The latter constraint does not apply, of course, when the person is not a competent adult. We are fully justified to practice paternalism when the person is a child or a profoundly retarded person, for example.

One moral argument for paternalism involves the notion of implied consent. In its simplest case, an unconscious accident victim cannot be asked for consent to a life-saving operation. We simply assume that if the victim had been conscious, he or she would have consented to the operation. This notion can be extended to include people without specialized knowledge, or people with superstitions that prevent actions clearly to their benefit. To use John Stuart Mill's famous example, if you know a bridge is to going to fall down, you have moral justification and obligation to prevent a person from walking on the bridge, even though the person might want to walk across in spite of your warning. By licensing professional engineers, society gives them a form of implied consent to do what is in the best interest of the public. For example, if the engineer closes a bridge because of imminent structural collapse, society agrees to prevent people from using the bridge.

The greatest difficulty with the first canon is that it assumes that there is a homogenous public out there that can agree on what is in its best interest. Obviously, in a pluralistic society, each person has an opinion and often these opinions differ markedly. In the case of the B. Everett Jordan Dam, a $120 million project in North Carolina, a large fraction of the university and professional society adamantly opposed the construction of what they viewed as a useless and destructive lake, while most of the farmers and local businesses favored what they saw as flood control and enhanced land values. The Corps of Engineers had to respond to many constituencies, each of whom had a voice and agenda, and often differed with one another. Engineers faced with such dilemmas should understand that the best they can do is to listen to all voices, present alternatives in as clear a manner as possible, and allow society through its elected officials to decide what is to be done. Engineering paternalism in the absence of public debate is normally not morally justified.

I was forced to do it. One of the hallmarks of a profession is autonomy. Professionals, theoretically, are free to choose which jobs they want to do and which ones they want to pass up. This is within limits, of course. A physician is morally obligated to assist an accident victim if by doing so he or she can save the person's life. Engineering seldom involves such life or death circumstances, and so engineers have greater freedom to do what they want [9].

But the ability to choose to not participate in a project assumes that there is little if any cost involved to the engineer. Because a majority of engineers are employed by private corporations or for the government, the level of autonomy is often limited. If the boss says to do something, the engineer is required to do it. This requirement is only applicable so long as the assignment is morally acceptable. Engineers are not required to perform assignments that involve immoral or illegal behavior or that result in the public's loss of safety, health, or welfare. More strongly put, engineers are obligated to oppose such actions. The excuse that the first canon was not adhered to because the government or corporation required otherwise is morally indefensible in a free society.

4.6 Deciding What, All Things Considered, Is the Right Thing to Do

Engineering solutions are predictable. A mechanical engineer designing an automobile can use the heat transfer equations in calculations with little concern for their accuracy of applicability. If there is a difference in temperature, the heat will be transferred from the hot side to the cold side, and the rate of that transfer is governed by the properties of the material. This will always work. No exceptions. If 100 engineers did the calculation, 100 of them would get the same answer.

But that is not how ethics works. If 100 engineers were confronted by a moral dilemma, one would expect at least 100 suggested solutions (and possibly a lot more if some people waffle). All that can be said is that some of these solutions are better than others. None is wrong, and none is right. Just some are better than others. So the question is, how can we design a system for coming up with answers to engineering ethical problems that will more often than not identify alternatives that most of us will consider acceptable?

Here is a suggested rubric for making an ethical decision in engineering:

1. *Find out what the relevant facts are.*
   Some problems disappear when the facts are all in. Getting accurate information can also avoid grave embarrassment.
2. *Determine what the moral issues are.*
   What exactly is bothering you? What wrong has been done or may be done? Is this a problem in engineering ethics, or is this a question of personal morality? If it is engineering ethics, is it a breach of the engineering code of ethics, or something more complex?
3. *Who is affected by the decision you have to make?*
   Include your own family, your friends, and others who will be affected by your final decision.
4. *What are your alternatives?*
   Here is where you want to be creative and "think outside the box." Perhaps you can come up with some imaginative alternatives that will not harm anyone and will not compromise your own integrity.
5. *What are the expected outcomes of each possible action?*
   We cannot, of course, accurately predict what the future will hold and what people will do. What is important here is that you differentiate between those actions that will undoubtedly occur and those that may occur.

6. *What personal costs are associated with each possible action?*
   We all have an obligation to do the right thing, but this obligation is limited by the costs we might incur. For example, if a certain ethical action will more than likely result in your losing your job, this is a large cost, but it may be acceptable if the situation demands it. On the other hand, if the probable cost is the loss of your life, then most rational people will agree that this cost is too high except in highly unusual circumstances.

In the 1930s Stalin purged (killed or imprisoned) thousands of engineers for asking too many questions, and it became quite clear that engineers in Soviet Russia were expected to perform only their technical function and to not ask questions about right and wrong. In that environment it is understandable that engineers simply did as they were told. Questioning the moral ramifications of political decisions would have resulted in an unacceptable cost. This appears to have been the single most important reason for the explosion of the nuclear reactor at Chernobyl. The engineers on duty were told to perform some tests, and although they knew that this was very dangerous, they did as they were told, without protest. When the alarms sounded, the engineers simply turned them off because they were told to do so. The spreading radiation displaced over 300,000 people and caused an extra 4,000 cancer deaths [10].

7. *Given the issues, alternatives, and costs, where can you get some help in thinking through the problem?*
   Wisdom, the ability to truly understand something, comes with age, and older people are a valuable resource. In engineering, the first place to look for help is in the code of ethics. Often, the first canon can trump all other considerations. If an alternative does not hold paramount the safety, health, and welfare of the public, then there is a very good chance that it is not the right one.
8. *Considering the moral issues, practical constraints, possible costs, and expected outcomes, what action should be taken?*
   This is the bottom line. All things considered, what ought you do?

4.7 From Ethics to Moral Courage

Figuring out the right thing to do is simply an exercise in ethical reasoning. The essence of moral behavior is not, however, the ability to reason, but rather *doing* what is right. After all, what good is understanding about honesty and fairness and professional responsibility if there is no willingness to act, especially in circumstances where the right thing to do is difficult? Knowing the right answer and then not implementing the desired action is pointless and even personally harmful. But often the right thing to do involves personal adversity, and it takes courage to make it happen. This kind of courage is called *moral courage.*

Courage is a moral virtue that makes it possible to act on all the other moral virtues. It is one thing to believe that being truthful is the best way to behave, but it is quite another to actually tell the truth. The first is a mental exercise; the second is an action that requires courage.

We admire people who have courage. Think of the still unidentified man who stood in front of the tanks during the Tiananmen Square massacre in Beijing. Thousands of innocent people were killed by the Chinese army, and there was that lone man, in front of the tanks. What immense courage it must have taken to do that!

Or the also unidentified German soldier in the Netherlands who, during the Second World War, was ordered to execute some villagers by firing squad. He refused to do so and was himself executed on the spot. These were acts of supreme courage because it was quite clear that they would suffer great personal harm for standing up for what they believed to be morally right.

The courage these men exhibited was not physical courage. Their courage was quite unlike the courage required for skydiving, or for walking on the moon, or for driving a tank through Baghdad. Theirs was the courage of inner conviction, that sense of knowing what the right thing to do is – and then doing it. We call this *moral courage* in order to distinguish it from physical courage.

Rushworth Kidder, in his book on moral courage, defines it this way [11]:

> *Moral courage* is the quality of mind and spirit that enables one to face up to ethical dilemmas and moral wrongdoings firmly and confidently, without flinching or retreating.

Moral courage is all about doing – of putting moral decisions into action.

There are two aspects to moral courage. The first is choosing to do the right thing, the second is choosing *not* to do the wrong thing. Moral courage is knowing that doing the right thing is dangerous, and then doing it anyway. Or, as John Wayne reputedly said, "Courage is being scared to death, but saddling up anyway."

Moral courage is needed both in personal life as well as in professional engineering. But because of their special responsibility to society, engineers have special obligations that demand moral courage. In the performance of their duties they have an obligation to do the right thing, even if it means upsetting their boss, losing a job, or incurring some other cost. It is the *role of the engineer* that carries with it responsibilities that exceed those of a private citizen. Not only are engineers expected to act morally as people, but they also have to act morally in their role as engineers and to have the moral courage to carry through with what they know to be right.

4.8 The Good Engineer

Person-to-person interactions are governed by manners, morals, and law. Often the boundaries between these are unclear. In fact, think of it as a Venn diagram with manners, morals, and law overlapping with each other. Some actions may be legal but immoral. Discriminating on the basis of race is legal (affirmative action, for example) but this is immoral because equal rights and opportunities may be denied to some. Other actions can be simply bad manners, but legal and not immoral, such as not acknowledging a good deed. And others actions can be bad manners, illegal, and unethical all at once.

In such human interactions we assume that all humans are rational and can make independent decisions. That is, we can decide to be ill-mannered, or we can decide to do something illegal. For us to act immorally, however, is to say that either there is something very special about us that allows only us to behave in such a manner or we must allow everyone else to act in a similar manner. Few of us are special in ways that allow us such privileges. And if we agree that it is OK for everyone to be ill-mannered, immoral, or illegal, then we effectively destroy society.

We in North America are so accustomed to the principle of equal opportunity and justice for all that we often forget that these principles are not universal. In some Islamic cultures, for example, women have only some of the rights and privileges enjoyed by men. Slavery is still practiced for profit in some societies. To be a gypsy in Eastern Europe today is to be a social outcast.

But we reject these alternative societal models and agree that we would not want to live in such environments. We have strong beliefs in the sanctity of humans and the worth of all people and do everything we can domestically and internationally to help others achieve the freedoms and privileges that we enjoy.

The foundation of good manners, high moral standards, and legality has been the argument that given the choice, we all would want to live in a society where everyone agrees to abide by the same manners, moral standards, and laws, recognizing that this benefits everyone. This argument is applicable to professional engineering as well. In order to help maintain a viable engineering profession, engineers need to demonstrate good professional manners and, if the occasion requires, admonish others for boorish behavior. They should act as role models in conducting engineering on a high moral level and promote such behavior in others. And without doubt they should not become criminals. In short, engineers all have a responsibility to uphold the honor of professional engineering and to create a culture in which all can flourish.

But there is a larger question of why any engineers should *act* in such a way. That is, if having bad manners, acting immorally, or even breaking a law is advantageous individually, why should we, at any given moment, not *act* in a manner that we would not necessarily want others to emulate?

The answer to that question is that we do not want to get caught and suffer the consequences. Bad manners would subject us to ridicule; unethical conduct might cause us to be ostracized by others; or we might lose clients and business. Breaking a law might result in a fine or jail time. But consider now the possibility where we would not get caught and could not suffer any adverse consequences. We never know for sure that we will not get caught, but for the sake of argument, let's assume this extreme case. We have it all figured out, and it is simply impossible for us to get caught being ill-mannered, immoral, or illegal. Why should we, all things considered, still act as honorable engineers, especially if this might involve financial cost to us or in some other way cause us harm? (Harm can be defined in many ways, but the usual definition is that which we would prefer not to have happen to us, something that would cause us an unwanted loss of resources or freedom or well-being. Aristotle might define it as that which causes pain or unhappiness.) The answer comes in three parts.

First, we are all members of a larger community, in this case the engineering community, and we all benefit from this association. Acting in a manner that brings harm or discredit to this community cannot, in the long run, be beneficial to us. Granted, the destruction of professional engineering may be far in the future and our small antisocial act would not be enough to destroy the profession, but we, along with all our contemporaries, have an obligation to uphold the integrity of the profession, ultimately for our own good. The conclusion is that engineers should act honorably because the profession depends on them to act honorably.

Second, the antisocial act, even though we might get away with it, takes something out of us. Sissela Bok, in her book on lying, contemplates the decision to lie or not to lie in a circumstance in which the lie will result in greater good at little cost to the teller [12]. For example, she agrees that it is obligatory to tell a lie to save an innocent life, but points out that every time a lie is told, the teller is less of an honorable human being. There is, as it were, a reservoir of good in each human and this can be nibbled away, one justified lie at a time, until the person is incapable of differentiating between lying and being truthful. This would also be true for manners and legal acts. Every time we get away with something we reduce our own standing as an honorable human being, and engaging in untruthful engineering reduces our own standing as a professional.

Finally, the reason for not being antisocial, even assuming we could get away with it, is that eventually our conscience would not stand for it. Michael Pritchard argues that we all have a conscience that tells us the difference between right and wrong, and most of us, when we do antisocial things, *know* we are behaving badly and eventually regret such actions [13].

But if, by telling lies, I become a scoundrel, why would I believe this to be an undesirable result? If I ran an engineering practice where I habitually lied to clients (e.g., "The report is in the mail," when in fact it is still being prepared), why is this detrimental? Why is lying (if I continue to get away with it) necessarily a bad thing?

Most people are harmed by knowing that they have acted dishonorably. The harm is to their own conscience. The assumption at the beginning of this argument that we limit the discussion to dishonorable actions that will not bring us harm is in fact an impossible assumption to make. Engineers who behave without regard to manners, morals or laws will eventually cause harm to befall themselves. They *will* lose clients, and their untruths *will* cause their works to fail, and they *will* have a bad conscience bothering them. They will eventually think poorly of their own standing in the profession and will regret the selfish self-serving actions that may have been ill-mannered, immoral, or illegal.

So why be a good engineer? We might get caught if we don't, we have a common responsibility to the professional engineering community, we lose something of our own integrity when we behave badly, and we have a conscience. But what if, in the face of these arguments, we still are not convinced? I must admit that there appear to be no knock-down ethical arguments available to make us change the mind of a person set on behaving badly. We have the human option to act in any way we wish. But if we have bad manners, act immorally, or break laws, we are not behaving honorably, and eventually we will be harmed by our regrets for acting in this manner. That is, such behavior will always result in harm to us as professional engineers.

While the Viking society of northern Europe was in many ways cruel and crude, they had a very simple code of honor. Their goal was to live their life so that when they died, others would say, "He was a good man." The definition of what they meant by a "good man" might be quite different by contemporary standards, but the principle endures. If engineers live their professional engineering lives so as to uphold the exemplary values of engineering, the greatest professional honor they could receive would be to be remembered as a *good* engineer.

References

[1] Vesilind PA, Gunn AS (1998) Engineering, ethics, and the environment. Cambridge University Press, New York
[2] DeGeorge R (1981) Ethical responsibilities of engineers in large corporations: the Pinto case. Bus Profession Ethics J 1(1):1–14
[3] James GG (1988) In defense of whistleblowing. In: Callahan J (ed) Ethical issues in professional life. Oxford University Press, New York
[4] Gert B (1988) Morality. Oxford University Press, New York
[5] Hobbes T (2009) Leviathan (ed. Gaskin JCA). Oxford University Press, Oxford
[6] Rachels J (2003) The elements of moral philosophy. McGraw-Hill, New York
[7] Regenstein LG (1991) Replenish the Earth. Crossroad, New York
[8] Vesilind PA (1993) Why do engineers wear black hats? J Profession Issues Eng Educat Pract 119(1):1–7
[9] Florman S (1996) The existential pleasures of engineering, 2nd edn. St. Martin's, New York
[10] Medvedev G (1991) The truth about Chernobyl. Basic Books, New York
[11] Kidder R (2005) Moral courage. Morrow, New York
[12] Bok S (1978) Lying: moral choice in public and private life. Pantheon, New York
[13] Pritchard MS (1991) On being responsible. University of Kansas Press, Lawrence, KS

Chapter 5
Engineering and the Environment

The engineer's commitment to society, the topic of the previous chapter, implies that "society" means "the human society." The moral courage needed for an engineer to do the right thing means that his or her actions would benefit the (human) public. That in itself is a fairly new idea for engineers, having been incorporated into American engineering codes of ethics only in the 1970s, but today engineers are charged with an entirely new responsibility – commitment to the environment. How did this come to be, and what does a commitment to the environment mean to modern engineering?

5.1 Evolution of Environmental Engineering

At the turn of the last century, when the first great environmental engineers – William Sedgwick, Ellen Sparrow Richards, Karl Imhoff, among many others – first began to define what we today know as environmental engineering, the objective was clear. People were still dying of water-borne diseases such as typhoid, dysentery, and cholera, and environmental engineers took on the challenge of providing safe and pleasing water for all. And they were spectacularly successful. Better sand filters, both slow and rapid, made possible by improved methods of coagulation and flocculation, produced sparkling clear water, and chlorination finished the process of disinfection. In less than 50 years the incidence of water-borne disease in the USA and other advanced countries decreased to essentially zero. The driving force in all this was public health, and the objectives were clear and unambiguous. There was no need to consider the ethics in any of this. It was done for the benefit of the people who paid for it and everyone agreed that this was the right thing to do.

During the Second World War American industry was mobilized to produce as much as possible, and questions of waste disposal were seldom considered. Much of industrial waste was discharged into watercourses, and by the 1950s the condition of the streams and rivers in our country was abysmal. Domestic wastewater was similarly not treated and most wastewater was discharged without any treatment. Only a handful of communities had wastewater treatment plants and few of these worked very well. Some rivers were so polluted that they occasionally caught fire, such as the Cuyahoga River near Cleveland, which incinerated a pier, and some were so full of industrial waste, such as the Nashua River in Massachusetts, that one could walk across the water.

This was unacceptable to the American people, and we initiated a concerted effort to clean up our waterways. Environmental engineers conducted imaginative research, used this information to design new treatment systems, and built and operated these plants so as to remove the major contaminants. They were spectacularly successful. The streams, lakes, and rivers of the USA today are comparatively clean and clear, and fish have been returning to their old habitats. Nobody asked questions about why we should do this. It was for the benefit of all the people. It was a good thing, and did not need moral justification.

About the same time that we started to recognize the dismal state of our lakes and rivers, the condition of urban air in many cities became intolerable. In Pittsburgh, for example, professional people who wore coats and ties to work had to change their shirts in the middle of the day because the soot in the air soiled the shirt collars. In Los Angeles the mysterious orange smog that descended on the city every day was found to be from automobile exhaust, and we started to develop ways to combat it. Engineers invented new devices such as electrostatic precipitators that remove the soot and sulfur from industrial and power plant emissions. Catalytic converters were installed on cars and photochemical smog was brought under control. There was no question of what needed to be done to clean up our air and the reasons were unambiguous. We were solving public health problems, and this directly benefited the people of the USA.

In the 1970s we discovered that not only had the war decades produced water pollution, but huge quantities of wastes had been dumped on land and these were slowly seeping into the groundwater. The Love Canal incident (in which 21,000 tons of toxic waste was discovered buried beneath the neighborhood of Love Canal by Hooker Chemical) precipitated a public outcry, and the engineers responded. They figured out how to quantify the waste, analyze its toxicity, and provide alternative plans for cleaning up the sites. They were spectacularly successful, taking care of many of the most egregious waste sites and doing it with little ill effect to either the public or to the workers. There was no moral quibbling about this effort. It was the right thing to do and the public demanded that it be done.

So at the end of the 20th century, environmental engineers could look back with pride on what they had accomplished. Clean drinking water, clean rivers, clean air, and a program to clean up the waste dumping sites. And it was all for the public who paid for it and appreciated the work of the engineers.

In the 1990s a new idea emerged – sustainability. The term was popularized in a United Nations report that suggested that all nations use resources in such a manner as to make them available to future generations. With sustainability, the beneficiaries are not the people who are paying the bills for reducing pollution, but some unknown and uncertain future generations. Recycling steel, for example, does not benefit people living today, for recycled steel often costs more than steel made from iron ore, but recycling benefits people who have not even been born, or who may never be born, because it reduces the use of nonreplenishable resources. Reducing methane emissions from wastewater treatment plants is a good thing because it reduces global warming, but only for our progeny. Our generation will never see any benefit from reduced methane emissions, nor will we be adversely affected by global warming. When we take on projects that have sustainability components to them, we are not benefitting present people who are paying the bills, but rather future people who may or may not even exist. If sustainability has no direct benefit to present people who are paying for the work, then there needs to be some justification for it. Why should we do this? Why should we embrace sustainability? Why, indeed, should we worry about future generations?

These are not trivial questions. It is easy to brush them off and simply say that "Of *course* we have to worry about future generations, and if you don't, then you are some sort of antisocial ogre." But not everyone agrees that we ought to invest societal resources (read taxes) in order to work toward global sustainability, and since we have no strong arguments for why we ought to work toward sustainability, it is possible that they are right. They would argue that sustainability is good only if it does not detract from our present pursuit of pleasure and wealth. Our objective on earth as humans is, they argue, to ravage the earth and glean as much as we can of its riches, and we need not be concerned with possible future people. How can ethics help us argue with these people and to formulate strong arguments in favor of working toward global sustainability?

5.2 Morals and Ethics

As discussed in Chapter 4, morality is a set of rules that ought to govern how people treat each other. These rules are accepted by rational people because they recognize that doing so will be to their mutual benefit. While we might disagree on details, it is clear that we agree on the most important moral values, and these same values appear in almost all cultures and traditions. As previously discussed, ethical models are used for deciding how, all things considered, we ought to treat each other when moral values conflict. If all situations require us to test only one moral value, then life is simple. One could simply choose to lie or tell the truth, for example. But what happens when moral values conflict – where it is not possible to abide by several moral values concurrently? This is a moral *dilemma,* defined as a situation where moral rules conflict. For example, suppose an engineer is asked to reveal some confidential information in court. The moral values being juggled here are truthfulness and loyalty. By revealing the information (being truthful) the engineer reneges on the promise made to the client (breaks a promise). The essence of *ethics* is to help us make such decisions by thinking through the problem using ethical models that might suggest solutions to the dilemma. This all applies exclusively to human–human interactions. The social contract assumes reciprocity; the belief that we all are able and willing to treat each other (that is, each of us humans) with respect and caring. But when we talk of *environmental* ethics, reciprocity no longer exists. Animals, plants, and things are not moral agents and cannot behave either morally or immorally. Therefore, there are legitimate questions about what exactly is meant by environmental ethics and why engineers should care one whit about the environment.

There are two broad approaches to environmental ethics. The first reason for environmental ethics is based on the proposition that the nonhuman environment has *instrumental value*; it is useful and valuable to people, just as other desirable commodities such as air, water, freedom, health, and opportunity. The second form of environmental ethics is based on *empathy* – our sense of caring for nonhuman nature.

5.3 Environmental Ethics Based on Instrumental Value

Instrumental-value-based environmental ethical models hold that since it is necessary for humans to live in a healthy environment and to be able to enjoy the pleasures of life, it is necessary to avoid contaminating the water, polluting the air, or destroying natural beauty. Despoiling the environment is morally wrong because it is like stealing something from other people. Environmental quality is important because it is valuable to us humans.

This form of environmental ethic also holds that nonhuman creatures have utility and should be protected. We would not want to kill off the plains buffalo, for example, because they are beautiful and interesting creatures, and we enjoy looking at them. Another reason for not exterminating a species is that there is the possibility that it will somehow be useful in the future. An obscure plant or microbe might be essential in the future for medical research, and we should not deprive others of that benefit. Finally, it would be unethical to destroy the natural environment because so many people enjoy hiking in the woods or canoeing down rivers, and we should preserve these for *our* benefit.

While the instrumental-value-of-nature approach to environmental ethics has merit, it also has a number of problems. First, if instrumental value is accepted as the sole criterion for an environmental ethic, this argument should not prevent us from killing or torturing individual animals as long as it did not harm other people. Such a mandate is incompatible with our instinctive feelings about animals. We would condemn a person who causes unnecessary harm or pain to any animal, and many of us do what we can to prevent this.

Second, this notion creates a deep chasm between humans and the rest of nature, a chasm with which most people are uneasy. In the instrumental-value-of-nature approach to environmental ethics we become the masters of the world and can use its resources for our sole benefit without any consideration for the rights of other species or individual animals. Such thinking led to the 19th-century "rape of nature" in the USA, the effects of which are still with us. Henry David Thoreau, John Muir, and many others tried to express in words what many people felt about such destruction, and recognized that it is our alienation from nature, the notion that nature only has instrumental value, that is the problem. Clearly the valuation of nature only on an instrumental basis is an inadequate approach to explain our attitudes toward nature.

On the other hand, a lot of good has come from this form of environmental ethics. The Endangered Species Act, for example, is legislation that prevents the destruction of species, not because they have some intrinsic value but because they may someday be useful to us [1]. While the instrumental-value-of-nature approach to environmental ethics is inadequate, it has nevertheless been the basis for a great deal of environmental good.

5.4 Environmental Ethics Based on Empathy

The second broad category of approaches to environmental ethics is based on empathy. Empathy is a complex notion and is poorly understood in moral philosophy. The word "empathy" has an interesting beginning, originating from the German word *einfühlung*, which means the ability to project oneself into a work of art, like a painting. Psychologists in the early 1900s searched for a word that meant the projection of oneself into another person, and chose the German word, translated into English as empathy. The concept itself was known, such as the Native Americans' admonition to walk in another's moccasins, but it simply needed a construction. The early meaning of empathy was thus the ability to project oneself into another person, to imitate the emotions of that person by physical actions. For example, watching someone prick a finger would result in a visible wincing on the part of the observer because the observer would know how this feels. Some observers actually feel the pain, similar to the pain of the person having the finger pricked, although often not as intensely.

From that notion of empathy it was natural to move to more cognitive role taking, the imagining of the other person's thoughts and motives. From here, empathy began to be thought of as the response a person has for another's situation. Psychologists and educators, especially Jean Piaget, began to believe that empathy developed throughout childhood, beginning with the child's first notion of others who might be suffering personal stress [2, 3]. The child's growing cognitive sense eventually allows him or her to experience the stress in others. Because people are social animals, this understanding of the stress in others, according to psychologists, eventually leads to true compassion for others.

However, we know from empirical evidence that not all children develop empathy toward the suffering of others, even though they might see or even experience such suffering themselves. Another problem with this notion of empathy development is that some experiments have shown that a person's state of mind is very important in that person's ability to empathize. A person in a good mood tends to be more understanding of others. Small gifts or compliments apparently significantly increase the likelihood that a person will show empathy toward third parties. If this is true, then empathy is (at least partly) independent of the object of the empathy, and empathy becomes simply a characteristic of the individual [4].

Charles Morris [5] defines empathy as

> the arousal of an emotion in an observer that is a vicarious response to the other person's situation … Empathy depends not only on one's ability to identify someone else's emotions but also on one's capacity to put oneself in the other person's place and to experience an appropriate emotional response. Just as sensitivity to non-verbal cues increases with age, so does empathy: The cognitive and perceptual abilities required for empathy develop as a child matures.

Such definitions of empathy are used in moral psychology, but they present some serious problems.

First, we have no way of knowing if the emotion triggered in the observer is an accurate representation of the stress in the subject. We presume that a pin prick would be felt in a similar way because we have had this done to us and we know what it feels like. But what about the stress caused by a broken promise? How can an observer know that he or she is on the same wavelength as the subject when the stress is emotional?

If a subject reports being sad, the observer would know what it is like to be sad, and would share in the sadness. That is, the observer would empathize with the subject's sadness and be able to tell the subject what is being felt. But is the observer really feeling what the subject is feeling? There is no definitive way to measure "sadness," for example, so there is no way to prove that the observer is actually feeling the same sadness that the subject is feeling [4]. There is, therefore, a built-in reporting error with empathy.

Most people have empathy toward animals in pain. Animals feel pain because their nervous systems resemble ours, and because their reactions to painful experiences are similar to ours. A sense of empathy can and does change the way people live. For example, many people do not eat veal because of the pain endured by the small calf, which, in order to become veal, is removed from his mother, fed liquid diet, and kept in confinement.

It is easy to feel empathy toward creatures that clearly feel pain, but what of the lower animals and plants? There is some evidence that trees respond physiologically when they are damaged, but the response may not be pain at all, but some other sensation (if we can even suggest that trees have sensations). Yet many of us are loathe cutting down a tree, believing that the tree ought to be respected for what it is, a center of life. This idea is best articulated by Albert Schweitzer in his discussions on the "reverence for life," or the idea that all life is sacred.

So our empathy toward the nonhuman world cannot be based solely on sentience. Something else is going on. When a person does not want to cut down a tree because of caring for the tree, this is certainly some form of empathy, but it does not come close to the definitions used by psychologists.

Another problem with this definition of empathy is that there is a huge disconnect between *empathy* and *sympathy*. If an observer watches a subject getting a finger pricked, the observer may know exactly what it feels like, having had a similar experience in the past. The observer may feel empathy, but there might be little sympathy. The observer might actually be glad that the subject is being hurt, or it might be funny to the observer to watch the subject suffer.

Years ago on the popular television show Saturday Night Live there was an occasional bit where a clay figure, Mr. Bill, suffered all manner of horrible disasters and ended up being cut, mangled, crumbled, and squashed. Watching this may have elicited some empathy on the part of the observers, but there certainly was no sympathy for the destruction of the little clay man. It was meant to be funny.

We could argue that a lack of sympathy might also suggest a lack of empathy. How could someone empathize with another person getting a finger pricked, but find it humorous? Perhaps there is no real empathy, or perhaps we have conditioned ourselves to laugh when others are hurt as a defense mechanism. This behavior might somehow separate the violence from our own experience. Or we have become desensitized by video games and movies depicting destruction without regret, or continuous footage of foreign wars to which we feel little personal connection.

Empathy is not a moral value in the same way that loyalty, truthfulness, and honesty are moral values. One chooses to tell the truth or to lie in a particular circumstance, and a moral person will tell the truth (unless there is an overwhelming reason not to, such as to save a life). However, one does not choose to have or to not have empathy. One either has empathy or one does not. One either cares for those in need or one does not.

Although morality is not a religion, moral people nevertheless also have a core belief that living a virtuous life is the right thing to do. Still, other than arguing self-benefit and illustrating cases where immorality can lead to personal disaster, it is not possible to teach someone to be a morally good person. If a person chooses to be a scoundrel, then there is very little that any instruction in ethics can do about it, and we who teach ethics should not be held accountable for failing our mission.

In order to drive this point home, an analogy with religion might be appropriate. A student taking a course in some religion, say Buddhism, might be asked on an exam what the central tenets of this religion are, and how these might be contrasted with other religions. The test might include questions on Buddhist values and how a person who believes in this philosophy might act in given situations. Students taking this course might do very well on such an exam if they have been diligent in their readings and have taken part in the discussions, but at the end of the course they need not believe in Buddhism or act in concert with the Buddhist philosophy they now understand. Understanding facts about Buddhism does not make one a Buddhist. Similarly, understanding ethical reasoning does not make a person ethical.

The issue is even more complicated in the case of environmental ethics. There is no reciprocity, so the logical arguments for ethical behavior no longer apply.

Despotic behavior toward the nonhuman environment, if such behavior does not hurt other humans, is perfectly "ethical" according to classical ethics. Environmental ethics based on self-interest is effective, and certainly this should and can be taught. But when there is no argument for self-interest, only the painfully thin empathy argument can be used to try to convince others that he ought not willfully cause animals pain or in other ways wantonly despoil nature. Because empathy cannot be taught, it is unfair to expect our schools to produce environmental engineers and scientists who have an empathetic view of nature.

I used to include a section on environmental ethics in my professional ethics course, and one of the activities was a video of a competitive coyote hunt in Wyoming [6]. The point of the film is that there is no economic or environmental reason for shooting these amazing animals. It is just "sport." The collection of carcasses at the end of the day's hunt is to me obscene and this was my reason for showing the film. I found that at the conclusion of the video there was an inevitable bifurcation of opinion. About half of the class believed that it was morally wrong to participate in such a "sport," and the other half of the class could see nothing wrong with it. These were just coyotes, for heaven's sake! Why would anyone care if we killed some of them? This difference of opinion was clear and deep. The students either did or did not care for the welfare of these animals. They either did or did not believe that people owed any moral consideration toward them. In short, they either did or did not have empathy for the coyotes. And if they did not, then there was nothing I could have said that would have changed their minds.

5.5 Sustainability

The concept of sustainability gained international recognition with the publication in 1987 of the report by the World Commission on Environment and Development, sponsored by the United Nations. Also known as the Brundtland Commission, their 1987 report, *Our Common Future,* introduced the term *sustainable development* and defined it as:

> Development that meets the needs of the present without compromising the ability of future generations to meet their own needs [7].

The underlying purpose of the commission was to help developing nations manage their resources, such as rain forests, without depleting these resources and making them unusable for future generations.

Sustainable development can be defined in a number of ways – and indeed the Brundtland Report itself includes ten different definitions – while a report for the UK Department of the Environment contains 13 pages of definitions [8]. Although the original purpose of this term was to recognize the rights of the developing nations in using their resources, sustainable development has gained a wider meaning and now includes educational needs and cultural activities, as well as health, justice, peace, and security [9]. Perhaps the engineers would have preferred "sustainable *management*" instead of development to describe the engineer's responsibility. For example, the New Zealand Resources Management Act, the comprehensive law that guides most environmental matters in New Zealand, refers to sustainable management instead of development as the national objective [10].

The concept of sustainable development has also resonated in the richer countries since it addresses intergenerational responsibility while acknowledging continuing technological change. Unstated was the assumption that intergenerational equity is a core ethic, and that future generations deserve as much opportunity to achieve a high quality of life as present generations.

If the process of sustainable development succeeds, it will lead to *sustainability,* or that stage of economic and technical development wherein the use of material and energy is at a steady state. The means for attaining sustainability is *green technology,* a term that recognizes that engineers and scientists are central to the practical application of the principles of sustainability to everyday life. The objective of green technology is to lead us to sustainability.

Even though it might seem clear enough on the surface, we still need to consider the morality of the search for sustainability. Should we promote sustainability and work toward achieving it? This is not a trivial question and it requires justification.

5.5.1 Can Sustainability Be Achieved?

Nature is a marvelous recycling machine. As an early environmentalist, Aldo Leopold, so eloquently described, a phosphorus atom in a plant becomes part of an animal when the plant is eaten and digested, and is liberated when the animal either dies or produces waste, dissolves in water, and is once again picked up by a growing plant [11]. Nothing goes to waste, and everything moves through the ecosystem in a logical and organized manner. Only energy is a one-way process, coming from the sun, picked up by the plants in the form of carbon–carbon and carbon–hydrogen bonds, used by animals, and finally by decomposers. The energy from the sun provides the power for the ecosystem to operate. If the energy from the sun is shut off, the global ecosystem ceases to exist.

The only creature on the face of the earth that deviates from this process is the human being. Is it possible to bring human activity into concert with the rest of nature? That is, can human activity become sustainable, just as natural ecosystems are? Sustainability is often described as the use of nonfossil fuel energy and the complete recycling of all materials, just as the natural ecosystems are able to do. Unfortunately, this is not an easy task for humans. Ayres has cataloged materials used in everyday industrial society with regard to the potential for recycling. Table 5.1 is a summary of some of his work.

Table 5.1 Potential for recycling of nonrenewable materials [12]

Class	Recycling possible?	Recycling economical?	Examples
I	Yes	Yes	Metals (aluminum, steel, copper), industrial catalysts, glass, corrugated cardboard
II	Yes	No	Solvents, refrigerants, plastics, automobiles, packaging materials (e.g., fast food containers)
III	No	No	Paints, pesticides, flocculants, explosives, fuels, detergents, fertilizers, lubricants, etc.

Some Class III materials such as fuels are chemically and physically altered so as to make them unusable, while some materials such as fertilizers are dispersed throughout the environment. As the level of dispersion of both the original and product materials increases, the amount of energy necessary to once again produce pure substances skyrockets. The point of this table is to demonstrate that there are many products that either would be extremely difficult to recycle or cannot under any circumstances be recycled. Given our present use of materials, the achievement of sustainability, if this is defined as perfect recycling, is simply not possible.

This conclusion can be derived analytically by using thermodynamics. The application of thermodynamic principles to economic models was introduced by Nicholas Georgescu-Roegen in his influential 1971 book *The Entropy Law and the Economic Process* [13]. The idea that materials' distribution in society can be expressed in terms of entropy, or the state of disorder, was first suggested by Berry [14]. Faber *et al.* developed an equation to calculate the energy requirements needed to produce a pure substance when its initial concentration is some mole fraction of the total [15]:

$$\Delta E = -RT\left[\ln x_i + \frac{1-x_i}{x_i}\ln(1-x_i)\right],$$

where ΔE = mole-specific energy requirement (W s/mol),
x_i = resource concentration (mole fraction),
R = ideal gas constant,
T = temperature (°K).

This equation defines the energy needed to produce a pure material if the material is widely distributed, or as its concentration approaches zero.

This problem would exist for most materials in human society, and would be especially true for the substances listed as Class III in the above table. As long as the availability of energy is limited, the environmental cost of producing pure substances (reducing the entropy) will be enormous and this will prevent the recycling of many materials. Thus the use of many nonrenewable resources is inherently unsustainable and any attempts to close this loop will result in far greater damage than good.

The problems with recycling have been unwittingly demonstrated by a project sponsored by the US Environmental Protection Agency (EPA). In an effort to develop reasonable and rational recycling goals, the agency constructed a mathematical model using life cycle analysis that incorporated energy use and environmental cost of a product. These costs included the extraction of materials, the manufacturing costs of the products, and finally the costs of disposal. The hope was to show how short-circuiting this process using recycling would result in energy savings and beneficial environmental impact. The final report showed that as the recycled fraction of a material (e.g., aluminum, steel, glass) increases, the cost in energy and adverse environmental impact (measured as global warming gas emission) also increases. This effect is not linear, and for most materials, as the fraction recycled passes about 25%, the environmental and dollar costs greatly increase [16].

Much of this dramatic increase in dollar and environmental costs is due to the way materials are used in society, and it certainly would be thermodynamically possible to collect materials at much lower cost if the materials were used differently in the original products. The increase in cost with higher recycling rates is not difficult to understand. Using glass as an example, some fraction of glass is easy and inexpensive to recycle, such as the voluntary curbside programs in many communities. But such programs recover only a small fraction of the total glass available for recycling. If the next step is to recover glass bottles out of mixed refuse in a materials recovery facility (MRF), the cost per ton is far higher than glass from curbside collection. Even this recovery will not result in all glass being recovered, and any other program, such as collection of glass from street and highway cleanup, will cost even more. And finally, beverage bottles are not the only glass in refuse and extracting the small pieces of glass would be prohibitively expensive.

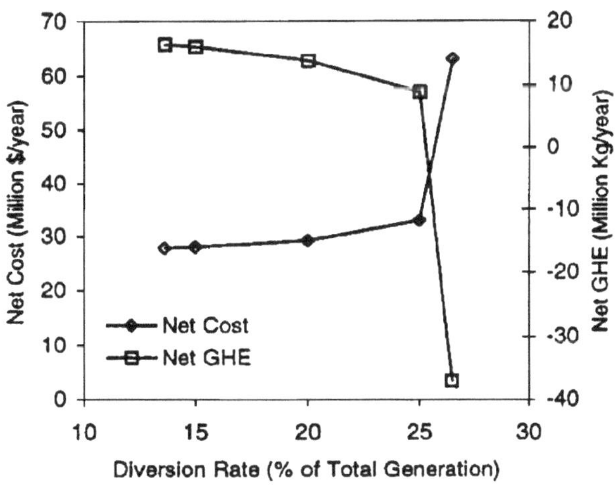

Fig. 5.1 As the rate of diversion from landfills (recycling) increases, the cost of recycling increases and the detrimental effect on the global environment increases. This figure shows that there is a dramatic break at about 25% recycling for glass bottles [16]

As the fraction captured increases, the unit cost has to increase. What the US EPA did not count on, however, was that the cost of recycling common materials such as glass jumped significantly at around 25%, a very low number. This finding caused some great concern within the EPA, where agency policy has been for many years to increase the rate of recycling, with 50% as a goal.

One could argue that it is necessary for long-term survival to use our energy resources to recycle materials. But such use will have adverse effects in many other ways, such as reduction in quality of life and changes in the function of natural ecosystems. If energy from the sun is captured and converted for human use, this energy is no longer useful for natural systems, which rely on the energy to produce complex organisms and systems. For example, the direct use of solar energy can result in temperature changes on land and in the water, and indirect uses such as the capture of wind energy can cause climatic changes. Even the capture of solar energy by photosynthesis (e.g., growing trees to produce fuel) reduces the availability of that energy for all other living matter. There is only a finite amount of energy available from the sun, and as this is diverted for use by humans to achieve order (reduce entropy), an equal amount of energy is no longer available to the natural systems. As Huesemann correctly observes [17]:

> The second law of thermodynamics dictates that it is impossible to avoid environmental impacts (disorder) when diverting solar energy for human purposes.

Therefore, the notion of full recycling/recovery of materials is essentially impossible given our present economic system. Even if we devoted our limited energy resources to greatly enhancing recycling, it is likely that the effects will not be what we desire. The use of energy for materials recycling could well result in a higher rate of the extinction of species, thus destroying the very thing we are trying to save.

If 100% is not possible, perhaps it is worthwhile to do what we can. Historically, humans have been exceedingly clever in the substitution of materials as some resources became scarce or too expensive. By this means, we have not seen the collapse of an economy as some useful materials have become unavailable. We have, in effect, achieved "sustainability" if this is defined as the continuous manipulation of materials in order to achieve some optimum level of use. Precious metals, for example, have been replaced by alloys in electronics, and organic lubricants have eliminated the use of lead in petroleum products. We would not have enough butter to go around if we had not invented margarine made from sunflower oils.

The continuous replacement of materials when the need requires it, that is, they become either too expensive or unavailable (including for environmental reasons), might be labeled *soft sustainability,* a kind of sustainability where we acknowledge that not everything can possibly be recycled but that materials can be substituted as needed. By contrast, *hard sustainability* is what occurs in nature, where material substitution in an ecosystem is rare. Hard sustainability requires the complete elimination of human effect in the dispersion of materials or use of nonreplenishable resources. This is impossible, as I argue above, and any attempt to use enough energy to achieve such recycling will result in far greater harm than good to the global ecosystem. Soft sustainability is both possible and feasible, and will remain so as our societies develop and move increasingly toward the goal of hard sustainability. The fact that we cannot today achieve hard sustainability ought not to dissuade us from practicing and reaching for soft sustainability.

An example of successful soft sustainability can be found on an Amish farm [18]. The farm, or the systems of farms in an Amish community, seems to be essentially sustainable, especially in the Old Amish order where no modern conveniences are allowed. Manure from cattle and horses, containing nutrients, is spread on land used to grow crops that then feed the animals. Human waste is taken care of on land and there is no waste produced that requires off-farm disposal.

5.5.2 Morality and Sustainability

Ethics is about how we ought to treat each other, and morality requires that there be reciprocity between people. I do not lie to you because I do not want you to lie to me. Just saying that lying is wrong is not a strong argument. We have to show *why* lying is wrong, and this we can do. All we have to do is to imagine what the world would be like if we did not agree to tell the truth. We could not believe anyone, and all information would be suspect. It would be a terrible condition that no one would want, and hence we have to agree that telling the truth is the right thing to do.

The same applies for environmental issues. I agree to pay to clean up a river because you might want to fish there and I do not want to deprive you of that pleasure. And I expect you to do the same for me. I do not litter the roadside, and I hope that you agree to not litter. We get along, and how we get along spells out our ethics. We have strong ethical arguments that dictate how we ought to treat each other, and a major element in these arguments is reciprocity.

With the concept of sustainability, however, reciprocity is impossible. We would not be doing things that benefit us, or even our children, and our actions may only benefit imaginary people who may never exist. How do we know what these imaginary people would want us to do for them? And since they don't exist (and may never exist), how can we have a moral obligation toward them.

The first challenge is easy to counter. Even though we do not know exactly who future people will be, we can be fairly certain that they will *not* want polluted streams, dead oceans, or poisoned land. We might be wrong on the details, but we are fairly certain about the essential issues. We know that putting too much lead or cadmium on farmland will be detrimental to future generations, just as we know that the legacy of hazardous waste sites, created before many of us were born, is detrimental to our generation.

The second argument is a more difficult matter. Do we owe moral commitments to people who do not exist? The best example that I know is to argue that a terrorist placing a bomb in a school is morally wrong regardless of how long the fuse on the bomb might be. Even if the bomb would not go off for 20 years, long after the present children have left the school, the act is still morally reprehensible. The conclusion is that it is morally right for us to try to achieve what we can in global sustainability, knowing full well that we cannot attain perfection.

5.5.3 Sustainability and Engineering Codes of Ethics [19]

Professionals correctly believe that professional autonomy is beneficial to the welfare of the public. If the government starts telling physicians how to treat people, or preachers what to preach, or engineers how to build things, then the public loses. Accordingly, the professions have jealously guarded their autonomy in the name of the public good. The engineering profession recognizes that if engineering is to maintain its professional autonomy, then the public has to trust engineers, and it is very much to the advantage of engineering and the public at large to maintain this trust.

Such autonomy can of course be taken away by the state, as witnessed in nations having totalitarian governments such as the former Soviet Union, in which once-proud and independent engineers became tools of the communist government and had little say in technical decisions. The inability of the engineers to voice their concerns about projects that were counterproductive, wasteful in resources, and harmful to the public was in great part responsible for the eventual downfall of the Soviet empire [20]. Engineers therefore need to maintain public trust, and one way of doing this is to adhere to a code of ethics.

The evolution of engineering ethics codes is discussed in Chapter 4. Recall that one of the earliest codes of ethics in the United States was adopted in 1914 by the American Society of Civil Engineers (ASCE). Based in spirit on the original Code of Hammurabi, which was not translated until 1904, the 1914 ASCE Code addressed the interactions between engineers and their clients, and among engineers themselves [21]. Only in the 1963 revisions did the ASCE Code include statements about the engineer's responsibility to the general public, stating as a fundamental canon the engineer's responsibility for the health, safety, and welfare of the public.

For all its merits, the 1963 ASCE Code of Ethics was a collection of "dos" and "don'ts" concerning *human* interaction – engineer/engineer, engineer/client or engineer/public. The Code only grudgingly and ambiguously recognized the involvement of engineers in environmental matters, long a controversial aspect of engineering. Responding to the general growth of environmental awareness, and conscious of the popular image of civil engineering as the perpetrators of environmental destruction, the ASCE Code was revised in 1977 to include the following statement:

> Engineers should be committed to improving the environment to enhance the quality of life.

Note first that "Engineers *should* ..." which is very different from "Engineers *shall*...". The use of "should" in effect precludes the enforcement of this section of the Code. All enforceable sections begins with the statement "Engineers shall ...". Further, note that environmental effects relate solely to quality of life. Although the Code is vague on the matter, the phrase "quality of life" presumably applies only to *human* life. The Code in no way suggests that nature has intrinsic value beyond its utility, or instrumental value, to humans.

Recognizing this deficiency in the Code, the Environmental Impact Analysis Research Council (EIARC) of the Technical Council on Research (a committee of the ASCE) proposed in 1983 an eighth fundamental canon [22]. The proposed canon reads:

> Engineers shall perform service in such a manner as to husband the world's resources and the natural and cultured environment for the benefit of present and future generations.

Listed under the canon are nine guidelines that elaborate on the canon. For example, guideline 8.g reads

> Engineers, while giving proper attention to the economic well-being of mankind and the need to provide for responsible human activity, *shall* [emphasis added] be concerned with the preservation of high quality, unique and rare natural systems and natural areas and *shall* [emphasis added] oppose or correct proposed actions which they consider, or which are considered by a reasonable consensus of recognized knowledgeable opinion, to be detrimental to those systems or areas.

The proposal struck many people as relatively modest and uncontroversial. It is explicitly anthropocentric: the environment is to be protected "for the benefit of present and future generations" – of humans, obviously. Nonetheless, in their January 1984 meeting, the Professional Activities Committee voted *unanimously* to *not* recommend approval, and the canon died there [23]. The reasons for this rejection noted in the minutes were that environmental concerns are adequately covered by Policy Statement No. 120. Members of EIARC, which first drafted this proposal, were told that legal considerations prompted the disapproval of the canon. A former chairman of the EIARC later admitted that this reason seemed implausible.

Some enlightened members of the ASCE decided to try an end-run, and created a different committee, the Committee on Engineering Responsibility (COER), and assigned it the task of getting the eighth canon approved. The committee met several times and was about to submit a revised canon for approval when suddenly the committee was disbanded. The appearance was that the ASCE leadership wanted to stop this movement and that disbanding the committee seemed to be the simplest way of rejecting the troublesome eighth canon without going on record as opposing it.

Many of the members and leaders of the ASCE continued to recognize the need for some definitive statement about the environment. Seeking alternatives to the eighth canon, the ASCE discovered the United Nations report on sustainable development. While this term seems to be an oxymoron (the earth cannot sustain limitless growth, so how can we sustain ceaseless development?), it allowed engineers to embrace development while still paying homage to environmental concerns.

From the start there has not been a clear definition of sustainable development, a situation that may have helped its popularization. Everyone can be for it and define it as desired.

Of all the options, the definition of sustainable development by the World Bank is both useful and popular [24]:

> The sustainable approach to development … contains a core ethic of intergenerational equity, along with an understanding that future generations are entitled to at least as good a quality of life as the present ones.

Most engineers would subscribe to this ideal. But engineers deal in operations – they are doing things – and therefore need an operational definition of sustainable development. Some have suggested a modified environmental impact analysis, taking into account not only the effect on the present environment but also considering effects on future generations and global ecosystems. But these techniques have the same problems as the original environmental impact studies – they depend on crossover valuation of often incompatible goods. What is more valuable, our present needs or the needs of future generations? How should the needs of wildlife for forests be balanced with the human need for lumber? What pain and suffering by a laboratory test animal do we accept in order to reduce health problems in humans? And who is to decide these questions?

Clearly the ideal concept is a long way from an operational definition. Still, with all its problems, sustainable development is a worthwhile idea. Recognizing this as a positive step in defining the responsibilities of engineers toward the environment, the first fundamental canon in the 1997 revisions of the ASCE Code of Ethics was changed to read:

> Engineers shall hold paramount the safety, health, and welfare of the public and shall strive to comply with the principles of sustainable development in the performance of their professional duties.

On the surface, ASCE has taken a giant step forward in incorporating environmental values into its Code of Ethics. But let us look at this more closely. Consider the wording. The engineer *shall* (that's a good start) *strive* (meaning that the engineer has to try, not actually do) *to comply with the principles of sustainable development*. But nowhere in the Code are the principles of sustainable development spelled out.

Principles of sustainable development are not like the laws of thermodynamics, regulations on stream quality, or traffic laws. Civil engineers wishing to practice in accordance with their society's Code of Ethics are apparently free to determine what, in their opinion, are the principles of sustainable development, and then all the Code asks of them is to strive to act so as to be in line with what they themselves determine to be these principles. By contrast, the Institute of Professional

Engineers in New Zealand and the Institution of Engineers in Australia have developed quite detailed policies on sustainable development (Australia) and sustainable management (New Zealand).

Unfortunately, the guidelines intended to support and explain the sustainable development clause do not offer much help. For example, paragraph 1f. reads:

> Engineers should be committed to improving the environment by adherence to the principles of sustainable development so as to enhance the quality of life of the general public.

Ignoring the curious reference to the "general" public (who else is there?), the key word is *should*. Even though the fundamental canon says *shall*, the guideline lets the engineer off the hook by suggesting that *should* is good enough.

The cynic might say that once again the American Society of Civil Engineers has changed its Code of Ethics to enhance its public image and not to effect a meaningful change in the actions of civil engineers. But this is unkind. Perhaps the engineers who drafted this revision were seriously trying to cope with environmental problems and balance the rights and benefits of humans against concerns for the nonhuman environment. There are many sensitive and caring civil engineers who want to do the right thing in the performance of their duties, and some of them no doubt were instrumental in effecting the change in the Code of Ethics. But as important as this first step is, the engineer who seeks guidance in the ASCE Code of Ethics for making decisions that affect the environment will be disappointed to find little of useful value in the Code.

Such ambiguity does not, however, excuse engineers from having concerns for the environment. It just makes it that much more difficult for engineers to do the right thing.

References

[1] 16 [1] 16 United States Code Annotated § 1531(a) (2009) Thomson/Westlaw, New York
[2] Piaget J (1952) The origins of intelligence in children. Norton, New York
[3] Piaget J (1960) The child's conception of the world. Adams, Littlefield, NJ
[4] Vaknin S (2005) Malignant self love: narcissism revisited. Lidija Rangelovska Narcissus Publications, Macedonia
[5] Morris C (1996) Psychology – an introduction, 9th edn. Prentice Hall, Englewood Cliffs, NJ
[6] High Plains Films (2002) *Killing Coyote*. Missoula, MT
[7] World Commission on Environment and Development (1987) Our common future (Brundtland Report). Oxford University Press, Oxford
[8] Pearce D, Markandja A, Baker E (1998) Blueprint for a green economy. Earthscan, London
[9] Herkert JR, Farrell A, Winebrake J (1996) Technology choice for sustainable development. IEEE Technol Soc Mag 15(2):12–20
[10] Gunn AS, McCallig C (1997) Environmental values and environmental law in New Zealand. Ethics Environ 2:23–26
[11] Leopold A (1949) A Sand County almanac. University Press, New York
[12] Ayres RU (1994) Industrial metabolism: theory and policy. In: Allenby BR, Richards DJ (eds) The greening of industrial systems. National Academy Press, Washington, DC

[13] Georgescu-Roegen N (1971) The entropy law and the economic process. Harvard University Press, Cambridge, MA
[14] Berry RS (1972) Recycling, thermodynamics and environmental thrift. Bull Atom Sci 28(5):8–15
[15] Faber M, Niemes H, Stepan G (1995) Entropy, environment, and resources. Springer, Berlin
[16] Solano E, Dumans RD, Harrison KW, Ranjithan SR, Barlaz MA, Brill ED (2002) Life-cycle-based solid waste management: II. Illustrative applications. J Environ Eng 128(10):993–1005
[17] Huesemann MH (2003) The limits of technological solutions to sustainable development. Clean Technol Environ Policy 5:21–34
[18] Kraybill DB (2001) The Amish and the state. Johns Hopkins University Press, Baltimore, MD
[19] Vesilind PA, Gunn AS (1998) Sustainability and the ASCE code of ethics. J Profession Issues Eng Educat Pract 124(3):72–74
[20] Graham L (1996) The ghost of the executed engineer. Harvard University Press, Cambridge, MA
[21] Hope RF (1994) The Code of Hammurabi, King of Babylonia. Gaunt, Holmes Beach, FL
[22] American Society of Civil Engineers, The Environmental Impact Analysis Research Council of the Technical Council on Research (1984) A proposed eighth fundamental canon for the ASCE code of ethics. J Profession Issues Eng 110(3):118–122
[23] American Society of Civil Engineers (1984) Minutes of the Professional Activities Committee of ASCE, January 1984
[24] Pezzy J (1992) Sustainable development concepts: an economic analysis. World Bank Development Paper No. 2, Washington, DC

Chapter 6
Peace and Justice

"Peace" is notoriously difficult to define. Unfortunately, the problem of defining what me mean by "peace" seems to have received little philosophical attention. A well-known encyclopedia of ethics, for example, does not have an entry for "peace." See "war and peace," it advises [1].

The word "peace" is supposedly derived from the Anglo-Norman *pas*, meaning "freedom from civil disorder," which in turn comes from the Latin *pax.* In Hebrew peace is *shalom* and in Arabic the greeting is *salaam.* This has a wider meaning and includes safety, welfare, prosperity, security, fortune, or friendliness. Peace can be defined positively as enjoying a healthy, secure, just, and free life. The negative definition of peace is the absence of war, and if we use this definition, we need to define what we mean by "war."

6.1 War

Every human culture has struggled with war. Some, such as the Viking culture, glorified war and saw nothing immoral about raiding and looting. Others, such as the Hopi, took defensive measures and built their homes on high cliffs so they could defend themselves from assault. As long as war, such as raiding neighboring villages, is "recreation," there is no need to justify it. Similarly, if the purpose of war is to gain riches and power, or to kill all those whose religion differs from your own because you believe your god wants you to do this, such wars are immoral and there is no need to even try to justify them. Such wars are started by immoral people who would scoff at any suggestion that one should have any concern for moral behavior. But if people who desire to live in a moral society and want to behave morally go to war, then there has to be moral justification for their actions.

Over many years, three approaches to warfare have developed in the West – the *just war theory*, *realism*, and *pacifism*. The first argues that in certain situations nation-states can be morally justified in going to war and that under certain circumstances it is ethically appropriate to use force. The challenge is to define the conditions that would make war morally just. Realism does not concern itself with morality at all. In this approach to war, the motivator is national security. Morality has nothing to do with war because the only thing that matters is survival. Finally, pacifism is based on moral thinking that leads to the conclusion that no war is just and therefore war is morally prohibited.

6.1.1 Just War Theory

The early Christian church had a problem. There was no equivocation about Jesus' teaching about harming other human beings. He was a pacifist and taught his followers to emulate him. However, once the Church had grown to the status of a wealthy nation-state, it needed to be able to protect itself from those who wanted to pillage Rome and steal its riches. Jesus would have given it all away, of course, but the Popes were not about to do that and needed a justification for defending Rome and the wealth of the Church against the Visigoths and other heathens.

The solution was found by Augustine (354–430), who developed what eventually became known as the just war theory [1]. He believed that war was necessary, but only under certain (Christian?) conditions. A conflict, according to Augustine, could be divided into three parts: *jus ad bellum, jus in bello*, and *jus post bellum*. In English, these are "before conflict", "during conflict," and "after conflict." Each has its own set of moral concerns and justifications.

The *jus ad bellum* argument defines when war is necessary and morally acceptable. If a just war is to be started, all six of the following have to be true:

1. There must be a just cause. One may attack only in self defense, to protect the innocent, or to punish wrongdoers. This does not apply to actions inside the nation state – only outside its borders – leaving open the question of aggressions within a country such as in a civil war.
2. The motivation has to be moral. Aggression is wrong if its purpose is to acquire more land or another's wealth. Revenge or ethnic hatred is specifically prohibited. The ultimate motivation has to be the end of conflict and the onset of peace.
3. The aggression must be approved by a proper authority, e.g., a legitimate government. This again leaves open the question of rebellion within a nation-state.
4. War must be the last resort when all alternatives have been exhausted.
5. There must be a high probability of success.
6. War must be waged with appropriate means. Weapons of mass destruction are not to be used if they are not needed.

It is instructive to set this list of necessary conditions against the beginning of contemporary wars such as the American invasion of Iraq. How many conditions were satisfied before American forces invaded Iraq?

Once war is under way, the *jus in bello* restrictions define how it is to be fought. There are three rules:

1. Only military targets may be attacked, although this restriction has seldom been adhered to in an all-out fight. During the Second World War, for example, having a red cross on a truck or a ship was no protection from attack. The German hospital ship *Moero*, steaming from Tallinn to Danzig with wounded soldiers and refugees, was clearly marked with red crosses and was still sunk by Russian aircraft. In the Pacific during the Second World War, US Navy corpsmen did not have the usual white dot and red cross painted on their helmets because the Japanese had been instructed to target them specifically so as to prevent wounded men from being rescued. The "Blitz" of London by the German Luftwaffe was nothing more than a civilian scare tactic. The Israeli incursions into Gaza have always resulted in civilian casualties and destruction of homes. As noted earlier, the bombing of Tokyo by General LeMay's B-29s killed over 100,000 civilians. Wars are replete with examples of nonadherence to this principle.
2. The force used has to be proportional to that used by the enemy. This restriction, related to old concepts of chivalry, is also typically ignored. The atomic bomb was certainly not in proportion to anything the Japanese had in their armory.
3. The use of illegitimate means (or *mal in se*, those weapons that are evil in themselves) is not allowed. This of course is wide open to interpretation. For many years the cross bow, which could pierce armor, was considered illegitimate and was supposedly banned. Today, we would ban, if we could, nuclear weapons, poison gas, biological warfare, mass rape, genocide, and ethnic cleansing, although weapons like land mines are still considered legitimate.

Finally, after the war, *jus post bellum*, there must be justice for war criminals who did not follow the above rules, but revenge by citizens is not permitted. The actions of the USA in Japan and in Europe are exemplary postconflict behavior. In Japan, the USA allowed the emperor to stay in office, which meant a great deal to the Japanese citizens. In Europe the Marshall Plan revived war-torn nations and led directly to the creation of democratic governments in western Europe. War criminals are, of course, identified by the winners. One could readily imagine a war crimes tribunal held by victorious Japanese after World War II. But again, the point of postconflict rules is to create a condition for peace.

This requirement has been brought forcefully to our attention in the Middle East. As the USA untangles itself from Iraq, it should leave behind a country with at least a chance to remain peaceful. Colonel Garland Williams writes about problems Americans faced in Iraq:

> Unlike conventional military operations, … peace operations [were] assigned to the military with little strategic political-military clarity. There [was] no "unconditional surrender" that can be demanded, signaling the end of the conflict and the end of American military engagement. Peace operations require a full analysis of the crisis situation in order to fully understand the totality of the problem and its symptoms, becoming the prerequisite for the United States to define the limits of its involvement [2].

The realization that American troops in modern regional warfare are responsible for more than shooting weapons has prompted a re-evaluation of the education of its officers. The curriculum at the US Military Academy at West Point is being reviewed and revised to respond to this need [3].

If war is necessary under the just war theory, then it is reasonable to conclude that a nation should do everything it can to win wars and that research on weapons systems and the development of the engines of war is a necessary part of today's national identity. If this is true, then engineers working on these projects are not necessarily doing anything immoral. Others would suggest that some types of military work are immoral because it cannot be justified under the just war argument. For example, the production of land mines might well be considered intrinsically immoral because of the harm unexploded land mines do to civilian populations. The use of "bunker buster" nuclear weapons would also be immoral if only one side in a conflict has such weapons.

Engineers have to recognize that the invention and implementation of such weapons are their direct responsibility. Such projects would not be possible without engineers. Other professions such as law, medicine, accounting, and so on are useful but not necessary to the development of armaments. Engineering is the only profession that is required in all instances where such development takes place. Engineers therefore have a responsibility that is far greater than that of other professions. A maxim of morality is that anyone who shares a necessary role in X has responsibility for X, and engineers have a necessary role in the development of modern weapons.

The just war theory approach to war is still strongly debated, with the most prominent thinkers being Christian theologians and philosophers [4, 5]. Apparently they are still trying to figure out what they would say to Jesus if he confronted them about their support of war.

6.1.2 Realism

"Get real!" is the old admonition of people arguing against something that simply does not match with reality. A realist would argue that in war, anything goes. The purpose is to win and not to try to be "nice." In fact, it would be immoral for a state, once at war, to try to act morally. They just need to win the war and worry about morality later. When confronted with losing, it is the state's responsibility to use whatever means it has to reverse the course of the war. Its only constraint is to behave in a way that prevents mutual destruction. A realist would argue that it is fine to follow rules, but only if the rules are to your advantage. The humane treatment of prisoners, for example, is only necessary because the other side might also mistreat its prisoners.

An example of a realist was former President Herbert Hoover, a mining engineer, who in 1954, long after he had left the White House, wrote a secret memorandum to President Eisenhower, warning him of the Cold War with the Soviet Union:

> It is now clear that we are facing an implacable enemy whose avowed objective is world domination. There are not rules in such a game. Hitherto accepted norms of human conduct do not apply. If the United States is to survive, long-standing American concepts of fair play must be reconsidered. We must learn to subvert, sabotage and destroy our enemies by more clever, more sophisticated, more effective methods than they use against us [6].

Much of modern warfare is fought under this banner, including the so-called "war on terror."

6.1.3 Pacifism

The pacifist believes that war is immoral, so one should not participate in war. This exposes the pacifist to the "free rider" criticism – enjoying benefits such as personal liberty without having to sacrifice to maintain them. If pacifists gained something from their actions, then this would be a powerful argument, but the reality is that pacifism, at least in today's environment, usually costs a great deal.

Pacifists are accused of not being in the "real world," of being idealists. Critics argue that if pacifists had their way, aggression would be rewarded and the world would be in the grip of bad people. In truth, in today's world little is to be gained by war. An aggressor cannot hold on long to a conquered people and eventually loses control. The best example we have of this is the history of eastern Europe, which was under the control of Soviet Communism for over 50 years, but which eventually overthrew its aggressor with little loss of life. In 1968, during the "spring of democracy" in Czechoslovakia, when the possibility arose for establishing a self-governing democracy, the Soviet Union sent in tanks and squashed the rebellion. The Czechs decided not to fight, much to the dismay of many in the West. But their decision was prudent. A few decades later they achieved their ends and became a free country without firing a shot. The Czechs remembered the 1956 revolt in Hungary where thousands died and for no good reason.

But the argument continues. What if the aggressor is brutal, like the Nazis in Europe and the Japanese in China? This criticism is far more difficult to counter because pacifism works only if the opponent has high moral values. Ghandi's pacifism, for example, worked only because the British had strong moral concerns about not harming innocent people. If the Brits had been like some aggressive countries in the not-too-distant past, they would have assassinated Ghandi and that would have been the end of the rebellion. We can play mind games and wonder what would have happened if the Iraqis had decided not to oppose the American invasion. We would have gone in, and right out again, with no loss of life, and the Iraqi people would still have a functioning country. The invasion of France by the immoral Nazi regime, on the other hand, was cruel and oppressive.

Only hardened pacifists would argue that this would have been the right response to the German invasion.

Pacifists argue for their beliefs from three separate perspectives. The first is the *religious pacifist* who believes that God commands nonaggression. True Christians, for example, would fall under this category, exemplified by the New Testament admonition to "turn the other cheek." A *deontological pacifist*, on the other hand, believes that a strong moral argument exists that war is wrong and that participation in war is an immoral act. Finally, the *consequentialist pacifist* believes that the costs of war are always too great and that refusing to fight always results in the greatest good for everyone.

The strength of any belief is measured by how persuasive it is to others. The first basis for pacifism, that God commands it, is not persuasive at all to someone who does not believe in God, or if God exists, that he (or she) has anything to say about war. The deontological argument is also difficult to express forcefully because it requires shared values. We must, for example, all believe that human life is valuable and sacred and that it is wrong to hurt innocent people. If these values are not shared, then the deontological argument has little impact in convincing others that pacifism is correct. The final argument is the most forceful if it can be made using empirical evidence. Can we, for example, find a pair of nations of a similar character, one of which chose to go to war while the other one chose not to fight?

One such pair might be the USA and Canada. In 1776, the lower colonies decided to rebel and fight the English, while the upper colonies decided to remain loyal to England. Now, more than 200 years later, we can ask if fighting the American Revolutionary War was truly the best option for the lower colonies. Canada enjoys the same freedoms and the same quality of life as those of us who live in the USA. While Americans glorify and venerate the patriots who fought the Revolutionary War, can we claim unequivocally that this war was necessary? Would North America be one large country with the same freedoms and the same quality of life?

A similar argument might be made about the Civil War, where losses to both sides were staggering. How long would slavery have lasted if we had not fought the war? We were almost the last country in the world to still have slaves at that time, and the system of slavery was clearly untenable. Emancipation would have occurred as soon as people understood that slavery was morally bankrupt and that the economic system did not require slavery. The public would have demanded the end of slavery by political means and not a single shot might have been fired.

While we cannot know for sure what would have happened in 1863, there is strong evidence that even though white southerners a hundred years later, in 1963, were all depicted in the media as horrid, prejudiced morons, the vast majority of them actually accepted the end to the Jim Crow laws with equanimity, resignation, and even relief. In a few months' time their world dramatically changed with the integration of schools and public facilities, and even though there were those criminals who made the headlines by forcibly interfering with integration, the majority of Southerners accepted this change without complaint. And the hero was a pacifist named Martin Luther King, Jr.

6.2 Positive Peace

Defining peace as the absence of war is too easy because this is a negative definition: defining what it is not. We might call this "negative peace." Negative peace is the use of the horrors of war to argue against war. Such arguments are not trying to attain some condition, but rather to prevent something else. For example, a sports team might strive not to lose and thereby win, thus defining winning as not losing. Negative peace uses descriptions and gory pictures of death and destruction to try to stop present or future wars. Deterrence by presumed mutual destruction, such as the standoff during the Cold War, is an example of negative peace.

If there is negative peace, then there must also be "positive peace," which would be more than an absence of war, but this would be a proactive effort to establish social justice through equal opportunity, a fair distribution of power and resources, and equal protection and impartial enforcement of the law. We would include in this list the support of international laws, compliance with multilateral treaties, use of international courts, nonviolent resolution of disputes, and participation in international organizations, trade, and communication. Positive peace would establish social equality and justice, economic equity, and ecological balance, protecting citizens from attack, and meeting basic human needs. A key element in positive peace is the availability of legal means to settle differences nonviolently and prevent violence. Positive peace seeks to eliminate the root causes of war and injustice and consciously attempts to build a society that reflects these commitments. Positive peace is what peace engineering is all about!

Positive peace is not a new idea. Johan Galtung, a Norwegian sociologist and cofounder of the field of peace research, has written extensively on positive peace. He believes the following:

> Positive peace is more than the absence of violence; it is the presence of social justice through equal opportunity, a fair distribution of power and resources, equal protection and impartial enforcement of law [7].

Maine Quaker Gray Cox, in his book *The Ways of Peace: A Philosophy of Peace as Action,* argues that we must create alternative ways of conceptualizing our economic, legal, and political systems [8]. He defines peace as a "process of agreeing," a definition that rejects the prevailing conflict-centered paradigm. His philosophy is based on the Quaker process of "communal discernment," which assumes that a holistic understanding of truth will emerge if all stakeholders can enter a dialog.

The United Nations agrees that peace is more than the absence of war and has formulated a program that promotes a "culture of peace" [9]. The charter that defines the program spells out the culture of peace as a set of values, attitudes, traditions, and modes of behavior, including the rights to:

- life,
- sovereignty,
- human rights and fundamental freedoms,
- peaceful settlement of conflicts,
- developmental and environmental needs of present and future generations,
- development,
- equal rights and opportunities for women and men,
- freedom of expression, opinion, and information,
- dialog and understanding at all levels of society and among nations.

The culture of peace can be achieved if we eradicate poverty, reduce economic and social inequities, equitably manage national debt, provide an adequate food supply, strengthen democracy, promote gender equality, and "ensure environmental sustainability, including preservation and regeneration of the natural resource base."

The United Nations document places great emphasis on culture because, and as Stanford anthropologist Robert Textor points out, culture is universally found among human groups and is essential to a group's sustainability [10]. Culture defines us and allows us to understand what we could be. If the culture is one of peace, then conflict will be considered immoral, unnatural, and repugnant.

Our present culture greatly affects how we treat each other. For example, not too many years ago men in the USA believed that women should not be allowed to vote, to hold office, or to do many other things. That was the culture of that time. Today, it would be inconceivable to suggest that women are second-class citizens. Although gender inequalities still exist, there are few *legal* ways that women are treated differently from men. We simply take it for granted that women have equal rights. It is our *culture* that defines this belief. It follows that if we can change our culture to eliminate chattel slavery, give women equal rights, and demand that our government not torture political prisoners, then we should also be able to make warfare a thing of the past.

Engineers can do a lot to implement the culture of peace, for peace engineering is *positive* peace engineering – the proactive search for peace. But the engineering establishment has a long way to go. We recognize that, given our present world order (or disorder), it would be extremely foolish to assume that all engineers, or all people for that matter, will soon reject war and behave peacefully. War seems to be built into our genes, and we will not soon give it up. We believe that we must defend our country (and defend others) fighting when necessary. Engineers working either directly or indirectly for the military/industrial complex are therefore needed in today's world. In addition to these engineers, however, we value those engineers who take a proactive approach to promoting peace and justice. If enough engineers devote their lives and skills to the proactive search for peace, military engineering may one day become unnecessary. We recognize that this day may be well into the future, but this should not prevent us from striving to achieve this goal.

6.3 Engineering and Justice

All engineers work at the behest of someone who needs something done and is willing to pay for the engineer's knowledge or skill. Most engineering ends up improving our standard of living, but sometimes engineering projects can cause great harm to others. For example, if an engineer is hired to construct a dam at a location where a predictable flood can readily cause the dam to fail, resulting in death and destruction downstream, should the engineer agree to do the design even though the dam's owner might be quite willing to pay the engineer and to accept the risk of dam failure? Does the engineer have any responsibility to the people who live downstream of the proposed dam, or is the engineer's only responsibility to the person who pays for his or her knowledge? Stated in another way: Is the engineer being fair to the people who are not directly paying him or her for doing the engineering work? Are these people treated justly?

Engineering became a profession when engineers began to be concerned about being fair to all – when they began to realize that their skills ought to be used in the service of society and that they have a responsibility to the public in the application of their technological skills. The recognition that engineers have a responsibility to society led to the adoption of engineering codes of ethics that state unequivocally that the engineer shall hold paramount the health, safety, and welfare of the public.

The use of the word "public" in this statement implies that the engineer must, in the performance of his or her duties, treat all people with respect. That is, the engineer has to be a *fair* engineer if he or she is to properly conduct himself or herself as a professional [11]. Although most engineers would probably have never thought about being fair, I would like to argue that the idea of fairness is the underlying moral principle that defines professional engineering practice.

Fairness is a moral concept, but it is quite different from other moral rules such as telling the truth, causing pain, and so on. Fairness is thought to be a higher level value, a more sophisticated concept that underlies many of the normative ethical theories. It is also more complicated than more common moral rules and difficult to characterize objectively [12].

The idea of fairness as a moral vehicle for individual and professional ethics was not well developed until John Rawls wrote his hugely influential book, *A Theory of Justice,* in which he proposed that fairness is justice [13]. For Rawls, justice emerges when there is a fair compromise among members of a true community. If individuals are fairly situated and have the liberty to move and better their position by their own industry, justice results when they agree on a mutually beneficial arrangement.

There are many definitions of "fairness," but not all of them are useful to understanding how engineers ought to behave in their professional roles. For example, there is the problem of the "free rider," a person who uses the contributions of others in society to better his or her position but who does not incur any of the costs to society. A person who does not pay taxes for religious reasons still uses roads and public services for which others pay. Many would deem such actions as "unfair" since that person is taking social goods without contributing to social welfare.

Another meaning of "fair" is the occurrence of good or bad events beyond anyone's control. For example, a person whose trailer is destroyed by a tornado while other trailers in the vicinity are spared would call their misfortune "unfair," although there is nothing unfair (in moral terms) about a random event of nature. However, if the random occurrence is followed by a willful act, such as increasing the costs of needed supplies following a natural disaster, i.e., "gouging," such an act *would* be considered unfair by many.

A popular use of the word "fair" relates to how events beyond the control of society treat a person. For example, someone might come down with a debilitating disease such as multiple sclerosis, a neurological illness that strikes only young people. Contracting multiple sclerosis, while a tragedy for the contracting individual and his or her family and friends, is not a case of unfairness. It is a sad event, but it is not unfair. On the other hand, if human suffering and increased risk of illness is caused by premeditated human actions, such as decisions to release toxic pollutants into the environment, then such decisions *would* constitute unfairness.

We need a definition of fairness that separates such unfortunate events as the random onset of a disease or destruction of property by natural forces from those where humans are responsible.

One option is to define fairness as a lack of envy, or when no participant envies the lot of any other. This does not necessarily lead to fairness, however, since the claims of some people might be exaggerated. For example, suppose a farmer is retiring and wants to distribute his 300-acre farm among his three children. If the children are equal in all significant ways, the farmer would divide the farm into three 100-acre plots. But suppose one child claims to be a better farmer than the other two and insists that he therefore receive a larger share of the 300 acres. A second child might need 120 acres so he can sell the land for a new airport, and so he stakes a claim for the larger lot. A third, who has more children than the first two, claims to need a larger share because this plot will eventually have to be subdivided.

Are any of these claims sufficiently legitimate to change the initial distribution of 100 acres each? It is unlikely that a disinterested arbitration board would respect any of these claims or alter the 100/100/100 distribution. Each of the three progeny might go away unhappy, but the process has nevertheless resulted in a "fair" division of the land. Further, the quality of the acreage, not simply the amount, is also a component of a fair distribution.

Another problem with the "envy-free" approach to fairness is that it depends on each person having a similar personality. Suppose one of the three children in the above example is a generous person and would not object to the other siblings taking much more than their share. At the conclusion of the division, one child getting 60 acres and the other two sharing the remaining 240 acres might result in an envy-free division of the land, but this would be eminently unfair to the generous one. Therefore defining fairness as a lack of envy does not seem to be useful. And, at its worst, it can be a tool for unfair distributions.

Another way of describing fairness is to define what we mean by its opposite, or unfairness. Nicholas Rescher identifies three types of possibly valid claims of unfairness [12]:

1. *Inequity*. Giving people goods not in proportion to their claim. The opposite would be *equity*, or a condition where people's shares are proportional to their just and appropriate claims.

For example, suppose three consulting engineers perform engineering work for a municipal engineer and all of them submit their bills for the work performed. The municipal engineer recognizes there are insufficient funds to pay all the engineers, and decides not to pay one of the consultants. This is inequity, the unfair distribution of goods (money in this case). It would be unfair for the municipal engineer to pay only two of the engineers.

2. *Favoritism*. Irrelevant conditions, for example one's relations or one's religion, should have nothing to do with the situation or claim. The opposite of favoritism would be *impartiality*, the evenhanded distribution of goods without favoritism.

Continuing the same example, suppose one of the consulting engineers submitting the bill for work done is a member of the same social club as the municipal engineer, and because of this, the municipal engineer decides to increase the pay rate for his friend. Showing such favoritism is blatantly unfair. Membership in the social club is not a sufficient difference to allow the municipal engineer to overpay his friend.

3. *Nonuniformity*. "Equal treatment under the law" means that the law is to be applied to all people regardless of their status or wealth. The opposite is *uniformity*, or the uniform application of the rules.

Continuing the example, suppose the municipal engineer requests proposals from consultants for the design of a new sewer. He asks two of the three consulting firms to submit proposals by a certain date, but intentionally tells the third firm that the proposals are due a week later than the actual due date. This firm will then more than likely miss the submittal date and not be eligible for the job. Such action is non-uniform treatment, and therefore unfair.

Another way of thinking about fairness is to define it as treating equals equally, and unequals unequally. That is, people should be treated the same unless there are substantive reasons for treating them unequally. For example, if an engineer designs a sewer to run along the side of the road in front of several houses, each homeowner deserves the same care in how their lawns are treated and the grass replaced. If the engineer decides not to re-seed the lawn of one homeowner because the engineer does not like the color of the house, this is not a substantive reason for such unfair treatment. A problem with this definition is that it is sometimes difficult to judge when all participants in a project are equals. Perhaps there are substantive reasons, such as the presence of swampy soil, why one homeowner's lawn should not be seeded after construction. Engineers in the field often have to make such decisions.

Society, in order to function, occasionally imposes unequal treatment of some. For example, take a private firm seeking a consultant to advise them on the design of a wastewater treatment process. If the project is quite expensive, they might decide to hire the best consultant they can, one who will charge them a high per diem. Paying this money to the expert consultant is not unfair to the other consultants, even though it might seem to be unequal treatment. The expert consultant is not equal to all the other consultants and deserves to be paid at a higher rate. This is not inequity or unfairness even though it is inequality.

Equality is, however, an objective of the legal system. This does not mean unqualified equality. Some identifiable groups of people such as professionals are treated differently under the law. All licensed pharmacists, for example, are authorized to dispense drugs, while this activity would be illegal for non-professionals. All people in the category "pharmacists" then are being treated differently from other people. Unfairness occurs when a pharmacist, because of some irrelevant differences such as gender, religion, or shoe size, is not allowed to dispense drugs. Similarly, while we want to treat all people the same when they have committed a crime, this seldom happens. A first offender might receive a different sentence than a repeat offender for the same crime.

Equality in state actions is also important, in that goods distributed by the state (and goods taken by the state) might not be equal, but might be equitable. The progressive income tax requires that people with higher salaries pay more on a per person basis than low wage earners, and welfare recipients need to show that they are destitute before they can receive assistance. The important objective of fairness is that each person be treated equitably (but not necessarily equally) within the process. So a high-wage-earning woman ought not to have to pay more taxes than a high wage-earning man, all other things being equal.

Perhaps this is a useful definition of fairness in engineering:

> A *fair engineer* treats all people according to democratically accepted and morally defensible societal rules, and whenever these rules result in unequal treatment, there must be a legitimate and morally acceptable reason for this inequality.

Fair engineers, therefore, are those who, in the use of their technical skills, treat all people according to democratically accepted and morally defensible rules.

The way engineers can act in fair or unfair ways varies according to the type of job the engineer has and who the employer is. The two primary types of employment for engineers are in the civilian sector and in the military sector. The problems with being fair can be quite significant.

Engineers in the civilian sector either work directly with public agencies or with private firms. The more direct contact an engineer has with the public, the more likely he or she will be confronted by an opportunity for fair or unfair behavior. Perhaps the best way to describe this is to use a real-world example – the Orange County Landfill episode.

As described by Azar, the story begins in Chapel Hill, now a booming community, and once a quaint village hosting the University of North Carolina (s. Azar, 1998, The proposed Eubanks Road landfill: the ramifications of a broken promise, unpublished). During the 1960s Progressive era Chapel Hill organized the first truly integrated school system in North Carolina, carving out the central section of town in a way that essentially integrated all schools. This forward-looking liberal attitude carried through in the election of municipal officers, and it was no wonder that Chapel Hill was the first town in North Carolina to elect an African-American mayor, Howard Lee.

At that time the town was using a small landfill owned by the university for the disposal of its solid waste, but this landfill was rapidly running out of space and the university wanted to close it. In 1972 a search commenced for a new landfill site. Searches then were not nearly as intense as they are today, and the entire process was quite informal. The town council decided that it wanted to buy a piece of land to the north of town for the new landfill. This land seemed like a good choice since it was between Chapel Hill and Hillsborough, the county seat of Orange County, and within a short distance of Chapel Hill. It was also a convenient location for Carrboro, a small community next to Chapel Hill. There were no new housing developments near the proposed landfill site, and it was off a paved road, Eubanks Road, which would facilitate transport to the landfill.

There was, however, a vibrant African-American community, the Rogers Road neighborhood, that abutted the intended landfill area, and these people sought out Mayor Lee to expressed their dissatisfaction with the choice of a landfill site. The mayor talked them into accepting the decision, promising them that this would be the one and only landfill located near their neighborhood; and if they could endure this affront for ten years the landfill would then be made into a neighborhood park. Most importantly, he told them that the next Chapel Hill landfill would be located somewhere else and that their area would not become a permanent dumping site. The citizens of the Rogers Road neighborhood grudgingly accepted this deal and promise and then watched as the Orange County Regional Landfill was built near their community.

The site for the landfill was 202 acres, split into two sections by Eubanks Road, and abutting Duke Forest, a research and recreational facility owned by Duke University. On one side of the site was the Rogers Road neighborhood.

The landfill, which had no liner or any other pollution control measures, opened in 1972. The three communities contributing to the landfill, Chapel Hill, Carrboro, and Hillsborough, along with Orange County, formed a quasi-governmental body called the Landfill Owners Group (LOG) to operate the landfill. The LOG was comprised of elected officials from the four governmental bodies. One of the early actions by this group was to establish a sinking fund that would eventually pay for the expansion of this landfill or a new site when this became necessary.

When the population of Orange County exploded in the 1970s it became obvious that this landfill would not last very long and that a new landfill would be needed fairly soon. LOG, using money from tipping fees, purchased a 168 tract of land next to the existing landfill, called the Green Tract, with the apparent intent of using it when the original landfill became full, but without actually publicly declaring that this was the intended use for this land.

By the early 1980s, when the new landfill became necessary, the Green Tract was considered to be too small to be a long-term solution, and a need was apparent for a larger site. The four governmental agencies asked LOG to initiate proceedings to develop a new landfill, to be opened in the mid 1990s.

The LOG set up a landfill selection committee (LSC) to oversee the selection of the new landfill and asked Eddie Mann, a local respected banker and civic-minded citizen to chair the LSC. The LOG directed the LSC to seek technical help with the selection process, and it hired Joyce Engineering, a Virginia firm that had assisted other communities in the selection of landfills, to conduct the search.

After a study of Orange County, Joyce Engineering selected 16 locations as potential landfill sites, using criteria established by the LSC such as proximity to cities, distance from airports, and avoidance of environmentally sensitive areas. One of the 16 sites chosen by Joyce was the Green Tract, which became known as OC-3. At this point the engineers ignored the promise made to the Rogers Road neighborhood and did not even include in their public presentation the history of that promise.

Next they held public hearings for the purpose of culling the list of 16 down to a smaller list for final discussion. As the 16 sites were being considered, each was named to one of three categories: 1) to be considered further, 2) to be placed in reserve for possible consideration later, or 3) not to be considered further.

The public hearings were classical "Not in My Back Yard" (so called, "NIMBY") exercises. Neighbors who lived around their proposed sites hired lawyers and environmental scientists or sought the help of local lawyers, physicians, and engineers who then tried to persuade the LSC that their site was inappropriate. In other cases the members of the LSC themselves had a personal reason to eliminate a specific site from consideration.

Following these hearings, the LSC pared down the original 16 sites to 5, one of which still was the Green Tract. Because the former Chapel Hill mayor's promise to the Rogers Road neighborhood that future landfills would be located elsewhere was not included in the engineering report, it was never even mentioned at the public hearing.

When the LSC members were asked later about the promise, they argued that since Howard Lee did not represent Carrboro, Hillsborough, or Orange County the well-intentioned promise was not considered binding to the other governmental entities. In addition, although Lee acknowledged making this promise, this was never found on any written document. Further, the people who were least able to resist the backdoor expansion of the existing landfill, the Rogers Road neighborhood, were told that the promises made by elected officials were null and void because the new politicians could not be held to promises made by former officeholders.

One problem with the Green Tract was that it was too small to afford a long-term solution, a source of encouragement to the Rogers Road neighborhood. But this was all changed when, late in the process and well after the public hearings, a new possible area for the landfill was introduced. Named OC-17, this site abutted the existing landfill and the Rogers Road neighborhood, and included a large tract of land in Duke Forest.

The opponents of these two tracts, OC-3 (the original Green Tract) and OC-17 (the new Duke Forest area) began to fight the selection process, aided by many Chapel Hillians who saw the inequity in this process. The resisters packed the LSC committee meetings, printed T-shirts ("WE HAVE DONE OUR SHARE"), wrote letters to the newspaper, and did everything they could to keep the inevitable from happening.

In 1995 the LSC approved the selection of OC-3 and OC-17 as the new landfill, but suggested that some form of compensation be made to citizens in the Rogers Road neighborhood. The decision next went to the LOG for their consideration.

The vote in the LOG was 6-3 in favor of the selected site. Two of the negative votes were by the representatives from Carrboro. The town of Carrboro would not be directly affected by the location of the landfill in the Eubanks Road area, and thus Carrboro ought to have had a clear selfish motive for choosing this site. But the two Carrboro representatives on LOG, Mayor Mike Nelson and Alderwoman Jacquelyn Gist, based their negative votes on the promise made by Howard Lee to the Rogers Road neighborhood, and announced that they would fight the selection of this site.

Nevertheless, having been approved by the LOG, the decision next went to the four governmental bodies for approval. Chapel Hill, Hillsborough and Orange County approved the site with little debate. In the meeting of the Chapel Hill Town Council the previous promise by Mayor Howard Lee was not even brought up. But Mayor Nelson and Alderwoman Gist convinced the Carrboro council to delay the approval until compensation could be worked out in advance of the decision, citing the previous broken promises as loss of trust in politicians.

This delay by Carrboro allowed Duke University to marshal its forces and to hire appropriate lawyers and scientists to come to the defense of Duke Forest. The university trustees voted unanimously to fight the siting, and the president of Duke, Nan Keohane, wrote a strong letter to the LOG and the four governmental bodies threatening legal action if the land in Duke Forest was to be taken. Using his knowledge of the area, the manager of the Duke Forest quickly located areas with endangered species and several wetland locations, thus reducing the available acreage for the landfill. A historic African-American cemetery was discovered in the forest and placed on the protected National Registry, further reducing the availability of land. But Joyce Engineering, still working only for their client, found ways to redesign the landfill so as to accommodate these restrictions and to still use the major part of the tract for burial of solid waste. Demands for public hearings and more tests did not change the decision, and a year after the vote, OC-17 remained the first choice of the LOG and the three governments. The government of Carrboro was under increasing pressure to acquiesce.

Then, in 1997, Duke University announced that it had deeded a section of Duke Forest to the federal government for conducting experiments. The federal government now controlled this land and the fight was over. It took clever legal work, the effective battle fought by the citizens of the Rogers Road neighborhood, and the courage of Carrboro's Mayor Nelson and Alderwoman Gist to stop the landfill from being sited at a location where the people had already done their share.

The most important characters for us, however, are the engineers who worked for Joyce Engineering, the consulting firm hired to find the new landfill site. They knew very well that a promise had been made to the people who lived on Rogers Road, and they had an opportunity to do the right thing by not including the Green Tract as a potential site. They could have, simply by not listing this site, kept the promise to these people, but instead they proceeded as if they were ignorant of the controversy. They could even have been a champion for the under-represented people of Rogers Road and argued against their client's wishes, but they forgot who they were really working for. The whole incident was not a proud moment for the engineering professions and certainly not for the firm of Joyce Engineering.

References

[1] Becker LC, Becker CB (1992) Encyclopedia of ethics. Garland, New York

[2] Williams GH (2005) Engineering peace. United States Institute for Peace, Washington, DC

[3] Manous J (2005) Military engineering education: to promote peace not war. In: Vesilind PA (ed) Peace engineering: when personal values and engineering careers converge. Lakeshore, Woodsville, NH

[4] Ramsey P (2002) War and Christian ethics: classic and contemporary readings on the morality of war. Rowman & Littlefield, Lanham, MD

[5] Holmes AF (ed) (2005) The just war: force and political responsibility. Rowman & Littlefield, Lanham, MD

[6] Moyers B (1989) The secret government: the Constitution in crisis. Public Affairs Television, New York, quoted in B. Roth, The moral arguments against military research. In: Mitcham C, Siekewitz P (ed) Ethical issues associated with scientific and technological research for the military. Annals of the New York Academy of Sciences, New York

[7] Galtung J (1996) Peace by peaceful means: peace and conflict, development and civilization. International Peace Research Institute, Sage, Thousand Oaks, CA

[8] Cox G (1986) The ways of peace: a philosophy of peace as action. Paulist Press, Mahwah, NJ

[9] International Decade for the Culture of Peace (2001–2010) United Nations, New York, http://www.unac.org/peacecp/decade/

[10] Textor R (2006) Building a proactive pro-peace culture in engineering. In: Vesilind PA (ed) Peace engineering. Lakeshore, Woodsville, NH

[11] Vallero D, Vesilind PA (2007) Socially responsible engineering. Wiley, New York

[12] Rescher N (2002) Fairness: theory and practice of distributive justice. Transaction Publishers, New Brunswick, NJ

[13] Rawls J (2005) A theory of justice. Belknap, Cambridge, MA

Chapter 7
Peace Engineering

The military/industrial complex is so ubiquitous that some engineers might wonder if there are alternative careers in engineering. They might recognize that military engineering (in all its forms, including working for defense contractors and conducting research for the Department of Defense) is destined to be used for warfare, either defensive or offensive, and they are unsure if they want to participate in such work. These engineers are looking for alternatives that would allow them to use their skills in a positive and proactive way to promote peace.

Because of such moral concerns, some engineers have changed careers to reflect their own interpretation of what "public" means in the engineering code of ethics and have devoted their professional lives to the use of technology in the pursuit of peace. Perhaps in time peace engineering will evolve and mature and eventually take its proper place alongside military engineering and civilian engineering. Engineers, and especially young engineers, have to understand that they can *choose* to work for war or to work for peace. Frederic II is no longer here to chain them to their posts, and they can elect not to participate in furthering lethal technology but instead choose to devote their careers to peace engineering.

Peace engineering is a new concept. The emergence of peace engineering at this time in the history of our civilization can be explained by the convergence of two developments. First, most engineers in the first world, or the West, are well off financially. They are not worried about survival, or the survival of their families, and they can always find work. Such freedom liberates them from the burdens of engineers who, only a few generations ago, had little choice but to find work where they could. Today, with our wealth and social underpinnings, engineers have much more freedom to choose a job or a career. In addition, many engineers find that they can create second careers after retirement, unburdened by the need for financial reward.

The second development that drives the evolution of peace engineering is the emergence of a sense of professional responsibility by engineers. Recall that it was only recently, in the 1970s, that engineers acknowledged publicly their commitment to hold paramount the health, safety, and welfare of the public. The concept of sustainability was adopted as a professional goal only a few years ago, representing a sea change in engineering. In today's universities, engineering students cannot graduate without a strong dose of professional ethics, and professional societies continue to police their memberships, expelling those who have committed serious ethics violations. The National Society of Professional Engineers and the National Academy of Engineering have both established a program in engineering ethics. The sum of all this effort is the emergence of ethics as an integral part of engineering.

Combining these two elements – the wealth of our society and the freedom it provides, and the recognition that "doing the right thing" is integral to engineering practice – have made it possible for peace engineering to become an alternative career choice.

The engineers introduced below have made that choice. These men and women have devoted their time and even entire careers to reducing inequities, mitigating natural disasters, and seeking ways to reduce human conflicts. The most important point to learn from their stories is that peace engineering is possible, and that there are an infinite number of ways to express it.

7.1 Exemplars of Peace Engineering

Dennis Warner (1938–)

Fig. 7.1 Dennis Warner on the Palestinian side of the "Wall of Separation"

With an undergraduate civil engineering degree from the University of Illinois, Warner joined the Peace Corps. Upon returning to the USA he finished his PhD in water resources engineering at Stanford. He tried academic life for a while but found it stifling. He then set out on a career of international service, working with the World Health Organization and other international agencies.

Warner has made a habit of being present at the worst international emergencies imaginable, and of helping to rebuild the sanitation and water supply systems. Among places he has served are Ethiopia during the famine, the Middle East during the 1991 Gulf War, the 1994 Rwanda genocide, the international conflict in Kosovo, the 2001 earthquake in India, the Palestinian Intifada in Jerusalem, and now the Iraq War. He is ostensibly retired, but you would never know it. At the beginning of the Iraq war, he joined a Department of Defense team that went in immediately after the invasion to help get water supplies back online. He was deeply opposed to the war, but at the same time wanted to do what he could to reduce the human misery caused by it.

He has spent most of his professional life working in (or for) very poor countries on issues of rural water supply, sanitation and environmental health. His most important contribution has stemmed from the recognition that when people have an understanding of how to address their immediate needs, they gain control over their own lives. This empowers them to take on additional development tasks. An example of his work was in assisting the people of Rwanda in 1994. Gangs of Hutu militia had killed 800,000 Rwandans, mostly ethnic Tutsis, in an organized massacre. This prompted the retaliation invasion of Rwanda by Tutsi exiles, who quickly toppled the Hutu-controlled government and drove it and some 2,000,000 refugees into camps in Zaire and Tanzania. The invasion stopped the genocide but left essential infrastructure shattered, cholera raging in the refugee camps, and most Rwandans, both inside and outside the country, severely traumatized and lacking basic services. As chief of rural environmental health in WHO, Warner visited the refugee camps in Zaire to advise on water and sanitation interventions to combat the cholera outbreaks and then worked closely with UNICEF to help coordinate water and sanitation relief efforts in Rwanda. He returned to Rwanda several times to help restore water quality testing laboratories and to develop a program for exhuming the mass graves created during the genocide.

Arup SenGupta (1951–)

Fig. 7.2 Arup SenGupta

An estimated 80 to 100 million people in India and Bangladesh drink water containing toxic levels of arsenic. Symptoms can take years to develop and include skin ulcers, tumors, loss of fingers and toes, and cancer. In parts of Bangladesh and rural India arsenic is found in water in concentrations as high as thousands of parts per billion. To place this in perspective, the USA regulates arsenic in drinking water at only ten parts per billion [1].

Arup SenGupta developed a simple and inexpensive well-head unit that removes arsenic from well water. With help from Bengal Engineering College in India and from Water for People, a Colorado-based nonprofit organization, he has installed the unit in more than 100 village drinking wells near Howrah and Calcutta in India.

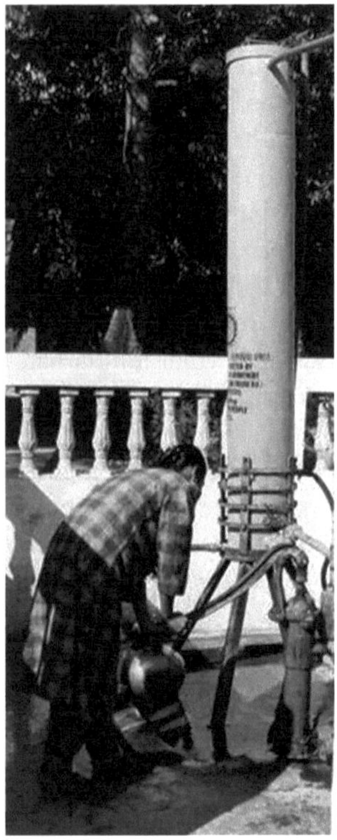

Fig. 7.3 SenGupta's arsenic removal column

SenGupta's filter is a green tube roughly 10 feet tall (Fig. 7.3). It attaches to the pump of the ground well and requires no electricity. When water travels through the tube, small beads containing iron oxide act as a filter to remove the arsenic. The filters are remarkably inexpensive and very easy to maintain. The filters are installed only in villages that welcome them, and they are carefully incorporated into the local communities. The filter operates by a hand pumping system with which the people have become familiar and therefore trust. The success of the filters is as much due to their appropriate technology as they are to the acceptance by the villagers.

Professor SenGupta grew up in Calcutta, India, and received his PhD from the University of Houston. There he focused on removing trace contaminants – an area of research he delved into even further when he arrived at Lehigh in 1985. In February 2007, the National Academy of Engineering awarded SenGupta the 2007 Grainger Challenge Prize Silver Award and $200,000 for his deceptively simple hybrid filtration system that is now helping thousands of people avoid arsenic poisoning.

Ben Linder (1959–1987)

Fig. 7.4 Ben Linder with friends

Occasionally a person has so much goodness, talent, warm-heartedness, and technical skill that it is difficult to capture the personality in a short paragraph. Such a person is Ben Linder.

Born in 1959, Linder studied mechanical engineering at the University of Washington and became an avid unicycle rider and designer. During his studies there he managed to design novel new mechanisms for unicyles that are now widely used. He also figured out how to install packs around the wheels of the unicycles in order to carry tents, sleeping bags, and other camping equipment. During graduation at the University of Washington, he smuggled his unicycle into the ceremony and rode across the stage to get his diploma, to the enthusiastic applause of the audience.

Ben loved life and cared for others, and this is why he decided to go to Nicaragua on his own to try to help some of the poorer people develop their resources. He was instrumental in setting up water supplies and hydroelectric projects near the village of San José de Bocay. This part of Nicaragua was unfortunately in the middle of the civil war raging between the Sandinistas, who sought to form a socialist government in Nicaragua, and the Contras, groups of mercenaries illegally funded and supplied by the CIA during the Reagan administration.

In 1987 the Contras attacked the small hydroelectric project, killing Ben Linder and his two Nicaraguan friends. The autopsy showed that Linder was first wounded by a grenade and then shot at point blank range in the head. This was no accident.

These types of projects were favorite targets for the Contras. Before Linden was killed, other Americans working to help the people of Nicaragua were murdered by the Contras, but the American embassy did not bother to even investigate these deaths. Marlin Fitzwater, the White House spokesman, was quoted in the *New York Times* as saying that Linder's death was his own fault, that anyone working in Nicaragua "put themselves in harm's way."

But most people understood what had been lost. Dan Rather of CBS News perhaps said it best:

> Benjamin Linder was no revolutionary firebrand, spewing rhetoric and itching to carry a rifle through the jungles of Central America. He was a slight, soft-spoken, thoughtful young man. When, at 23, he left the comfort and security of the United States for Nicaragua, he wasn't exactly sure what he would find. But he wanted to see Nicaragua first-hand, and so he headed off, armed with a new degree in engineering, and the energy and ideals of youth. This wasn't just another death in a war that has claimed thousands of Nicaraguans. This was an American who was killed with weapons paid for with American tax dollars. The bitter irony of Benjamin Linder's death is that he went to Nicaragua to build up what his own country's dollars paid to destroy – and ended up a victim of the destruction. The loss of Benjamin Linder is more than fodder in an angry political de-

bate. It is the loss of something that seems rare these days: a man with the courage to put his back behind his beliefs. It would have been very easy for this bright, young man to follow the path to a good job and a comfortable salary. Instead, he chose to follow the lead of his conscience [2].

Tarmo Soomere (1957–)

Fig. 7.5 Tarmo Soomere

Heroism is doing something above and beyond duty in the face of danger. Tarmo Soomere does not think of himself as a hero because he only did what he had been trained to do and what was expected of him. He is too modest, because his "above and beyond" saved the lives of countless people in Estonia.

Soomere is a coastal engineer based at the Institute of Cybernetics at the Technical University in Tallinn, Estonia. A few days before 8 January 2006 he looked at the weather data and could hardly believe his eyes. There was a "perfect storm" gathering in the Baltic Sea, and his instincts told him that this was bad news for the people living on the coast. A call from a colleague in Denmark confirmed his suspicions. The storm would be one of the worst ever, with wave heights over 10 m (about 30 feet). But how to warn the people who lived on the islands and on the coast? Estonia did not have anything like a marine early warning system, so there was nobody to tell of the impending disaster. At this point Soomere could have just closed his office and gone home for the night.

Luckily he knew something that could save lives and avert disaster, and instead of going home, he recorded a warning that he then sent electronically to local TV and radio stations. He stayed on the phone talking to journalists at the stations to make sure they took his warnings seriously. During the evening hours of 8 January his warning was broadcast many times over the radio and TV stations, and people got the message. They had experienced severe coastal storms before and knew what to do.

Soomere had succeeded in getting the public's attention, and they took the necessary precautions. It was a good thing they did, for the storm that struck the western coastline of Estonia during that night was one of the worst in decades. A number of harbors were destroyed, offshore islands were flooded, and many coastal residents were marooned as the cold Baltic water sloshed around their homes.

Because of Soomere's warnings, most of the people had had time to prepare for the storm. Small boats were pulled far onshore, and everything that could be tied down was lashed. Soomere's hard work in getting the message out saved countless lives. He became a folk hero to many, and the largest newspaper in Estonia, the *Postimees*, named him Man of the Year. The episode has prompted the government of Estonia to establish a marine early warning system, reducing the risk of damage and loss of life from future "perfect storms" in the Baltic Sea [3].

Marc Edwards (1964–)

Fig. 7.6 Marc Edwards

Lead is present in all natural waters and is not removed in drinking water treatment systems. These concentrations of lead are usually very low, far below any public health concern. But when drinking water travels through old lead pipes or through copper pipes that have been connected with lead solder, there is danger of the lead dissolving in the water if the chemistry of the water allows this to happen.

For years water treatment plants have been using free chlorine for disinfecting drinking water. Chlorine is, however, dangerous to water plant workers and, if improperly used, can produce residual organic chlorinated compounds shown to be carcinogenic. An alternative disinfectant is chloramines (compounds of chlorine and ammonia), which are equally effective disinfectants but are much safer to use and do not result in high concentrations of carcinogenic chlorine compounds.

The US EPA had recommended that water treatment plants start using chloramines, and many communities had switched over, including the District of Columbia Water and Sewer Authority.

For many years the Authority had been plagued by leaking pipes in homes. The copper pipes would develop tiny holes, and since the water is under pressure, the tiny streams of water would spurt out, causing water damage. One expert in the formation and prevention of pinhole leaks was Marc Edwards of Virginia Tech. He was funded by the Authority to conduct research intended to find out why leaks developed in copper tubing. Edwards knew that the type of disinfectant used had a major effect on the pinhole leak problem, and that the disinfectant also changed the chemistry of the water. But in conducting chemical analyses of the water samples, he discovered something both unexpected and troubling. The lead levels in Washington area drinking water were hundreds of times higher than what would have been considered acceptable. That meant that thousands of people, and especially children, were being poisoned by lead, which affects the nervous system and causes mental disability. He went to the DC Water and Sewer Authority with his findings, and their response was to cut off his research funding and to threaten him with other actions if he published his results. Their intent was to hush up the problem. But Edwards refused to buckle under what amounted to professional blackmail and went to the press. An article in the *Washington Post* blew the lid off the scandal.

Edwards, who earned his civil engineering degree from the University of Buffalo, showed that if chloramines are used as the disinfectant, the lead that the water picks up from pipes and solder does not precipitate but remains in solution, eventually traveling to the tap. He understood that the change in disinfectant must have caused excessive corrosion of lead pipes, lead solder, and leaded brass plumbing in homes, and this resulted in lead leaching into and contaminating the drinking water.

Edwards' findings resulted in congressional investigations, new laws to protect public health, and a critical Government Accountability Office investigation. More recently, the US EPA, which cited the water utility for violations, called the utility's practices unprecedented and a "serious breach" of the law. None of this would have happened if Marc Edwards had caved in to the demands of his funding agencies and stopped his research [4].

Fred Cuny (1944–1995)

Fig. 7.7 Fred Cuny

From an early age, Fred Cuny wanted to be a fighter pilot. He went to Texas A&M to study engineering with the intent of entering Marine aviation. But during his sophomore year he got involved in a prank that could have had serious consequences, and he was expelled. He enrolled in a smaller school in Texas and finished his engineering studies, going on to the University of Houston to get a master's in city planning. During school he was a strong conservative, but a science project that involved assisting migrant Mexican workers got him interested in their plight, and his politics changed dramatically.

On graduation, he still had hopes of a flying career, but a serious accident crushed one of his legs and ended his dream of being a fighter pilot. His first job was with an engineering consulting firm that specialized in sanitation problems. Cuny was sent to a small border town to investigate the high rate of illness there. He saw immediately that the most effective solution to the health problem was to pave roads and get rid of stagnant pools of water that acted as breeding grounds for insects. This was an inexpensive and imaginative solution, since most engineers would have recommended water plants and distribution systems.

Cuny discovered he had a knack for solving practical problems and, having studied African history in college, wondered if his skills might be useful to people affected by the Biafran-Nigerian war. He offered his services to Nigeria, but the Nigerian government did not want his help and kicked him out of the country. He next went to Biafra and personally witnessed devastation and hunger caused by the long war. He organized an airlift to bring food and medicines into the country, working with the NGOs. He also figured out that food distribution was the main problem and devised a system of sending food to villages, preventing the people from coming to the cities where sanitary conditions were extremely poor.

When he returned from Biafra, he founded an organization specializing in giving technical assistance and training in disaster relief to volunteer agencies. It was not easy making a living like that, and he barely survived. But disasters kept coming, and he was involved in many more during the 1970s, including Guatemala, Bangladesh, Cambodia, Thailand, and Calcutta. In Guatemala he showed villagers who had lost their homes to the devastating earthquake how to rebuild their homes using existing material and how to make new building blocks stronger than the old ones so that they could resist earthquakes. He went to Kuwait after the Gulf War and helped the Kuwaitis rebuild their water supply. In northern Iraq, he got Kurds to return to their homes and set up infrastructures so they could establish their own societies. His efforts were instrumental in the recovery of the Kurds after the Gulf War. In Somalia he worked on food distribution and then set up schools in Bosnia. In 1993 he went to Sarajevo and through various (perhaps not all legal) means was able to restore gas distribution to the city so that people could heat their homes and apartments. His most amazing feat was, during the siege of the city by the Serbs, constructing a complete water treatment plant in an abandoned tunnel, providing water to over 120,000 people.

Cuny recognized that too many relief efforts after natural or manmade disasters were poorly handled and managed, and noted that often the people on the ground were young and inexperienced. His idea was to create an educational opportunity for engineers interested in disaster relief.

Cuny lost his life in Chechnya, where he had gone to try to alleviate the conditions in the bombed-out country. He, along with two Russian doctors and an interpreter, were apparently executed by the Chechens after the Russians had spread lies about them, such as their being anti-Islam. After they were executed, the Chechens apparently realized that they had killed someone important and destroyed their remains. We will never know for sure what happened to Cuny, but his legacy as a "master of disaster" will long live in our memories [5, 6].

Peter Hagelstein (1955–)

Fig. 7.8 Peter Hagelstein

During the Reagan administration, the development of a missile shield over the USA, so-called Star Wars, became the single largest governmentally funded research project. Among those engaged in the work was engineer Peter Hagelstein, who was based at the Lawrence Livermore (California) National Laboratory, one of the world's largest military establishments. Hagelstein was critical to the mission because he had conceived the concept of the X-ray laser, which some thought would be the weapon most likely to be used as a defense against missiles.

The Soviet Union, against whom this defensive shield was intended to be employed, regarded Star Wars as a provocation. They did not share the view that the shield was a defensive weapon, but instead saw it as an offensive shield that would allow the USA the freedom to attack without fear of retribution. The inception of this program elevated tensions between the USA and the Soviet Union.

The second problem with the program was that most of the respected and disinterested scientists in the USA did not believe the system would work. They based their arguments on statistics – the incredibly small probability that a laser beam could actually find a missile flying at many times the speed of sound; these critics believed the system would be even less effective when there were thousands of missiles to be destroyed at the same time.

Some scientists and engineers enthusiastically embraced the concept of a missile shield and proceeded to try to make it work. Others were more skeptical but decided to put on blinders and compartmentalize their work and their moral principles. A notable and significant exception was Peter Hagelstein. He made an apparent decision of conscience and decided to resign his position at the Lawrence Livermore labs, even though he had invented the very weapon that was to be used for Star Wars. After resigning, he went to MIT, telling friends that he wanted to do research to "benefit all mankind." At MIT he continues his work with the X-ray laser, but now it is directed toward applications in medical research [7].

Jens Aage Hansen (1938–)

Fig. 7.9 Jens Aage Hansen

There are many ways for the more wealthy nations to reach out to less fortunate countries. Giving short-term emergency aid is certainly admirable and necessary, but more effective is the building of national infrastructure so that a country can become self-sufficient. Jens Hansen, a civil engineer at the University of Aalborg in northern Denmark, argues that research universities are critical in developing useful knowledge that can find immediate use in a developing country, resulting in economic and social success. He believes that sustainable development relies on an innovative society, one that can continually adapt to both global and regional challenges. The continued well-being of a society must be founded on a functioning national information system based at major universities.

Toward this end, Hansen and his colleagues at Danish universities have created an ambitious plan that supports and encourages developmental research and outreach at universities in many developing nations. The work is financed by the Danish International Development Assistance program as well as the World Bank. The Danish effort includes a number of separate programs (due to financial constraints), but each program assists universities in developing countries in the building of their research and development capacity. Universities participating in these programs include those from Malaysia, southern Africa, Thailand, Costa Rica, Nicaragua, El Salvador, and others. The Danish programs pair European universities with those in developing countries in order to exchange students and faculty and to sponsor international symposia. The hope is that the universities in the developing countries will join the global knowledge society and that the investment in capacity building in higher education will play a key role in this development.

Daniel A. Okun (1917–2009)

Fig. 7.10 Daniel Okun

The World Water Council estimates that presently one in every three people lacks adequate sanitation and one in six does not have a supply of safe drinking water [8]. This is an appalling statistic, and Dan Okun devoted his life to correcting it.

Okun first became interested in sanitary engineering when he watched his father work on the tunnel bringing water from the Delaware River to New York City. He went on to study engineering at the Cooper Union and received his doctorate at Harvard under Gordon Maskew Fair. After a stint in the army during the Second World War, he came to the University of North Carolina at Chapel Hill (UNC-CH), where he built an excellent department of environmental engineering. At UNC-CH he became a consultant to water managers all over the world and helped many countries develop programs for providing safe drinking water to their people.

Okun realized, however, that as populations grew, there was a severe lack of engineers capable of designing and constructing urban water supplies. The only way to get the job done was to bring engineers from developing countries to the USA, train them, and send them back to work on building secure water supplies. He received funds from various agencies to run a program called the International Program in Sanitary Engineering Design (IPSED) in which engineers from other countries came to UNC-CH to take courses that were especially designed for them and outside the realm of regular graduate program. After they completed their class work, they were placed in short-term work/experience activities with consulting firms, industry, and elsewhere. Okun was instrumental in setting up another program called Environmental Training and Management in Africa (ETMA) in which teams from UNC-CH and consulting firms conducted 2- or 3-week training programs in Africa, also with support from the US Agency for International Development.

The most important lesson the IPSED engineers learned in Dan Okun's programs was to *not* emulate the methods and materials used in the USA for water supply development, but rather to work with their own people and to adopt appropriate technology for solving environmental engineering problems. It is impossible to estimate the magnitude of the effect Okun's programs have had on the health and well-being of people in developing countries because knowledge gained by engineers who benefited from the training has been multiplied manyfold as they taught more engineers in their own countries [9].

David Schaad (1968–)

Fig. 7.11 David Schaad

When natural disasters strike, they can cause acute problems in hunger and disease, but also more sustained problems in the destruction of livelihood. Working through Engineers Without Borders (EWB), David Schaad and a group of civil engineering students from Duke University recently traveled to Lamnga, Indonesia, to a small village named Aceh where the 2004 tsunami had destroyed shrimp beds that sustained the village. The tsunami had killed more than 225,000 people in 11 countries, inundating coastal communities with waves up to 30 m high. The village had, during the ensuing year, rebuilt most housing, but shrimp beds that they needed for communal income had seen a severe drop in production. Part of the problem was that the water in the beds needed to be aerated and the only source of oxygen was the washing in of sea water. If they could increase the oxygen levels in the ponds, they could greatly increase the production of shrimp.

Schaad's team of students designed a mechanical aerator that did not depend on electrical power, which was fickle. Their aerator could be operated by hand and was much more dependable. They tested the aerator on a small pond on campus before they left for Indonesia. When they arrived there they were challenged to find the materials necessary to construct the aerator because it made no sense to construct it from components that were difficult if not impossible to obtain by the villagers. Their hope was that the device would be used and duplicated by the village once the team had left. Using the help and advice of the local villagers, the aerator they constructed worked well and the team was quite satisfied with its efforts.

Although the project was successful, the effect of the trip on the students was even more striking. All of the students were asked to keep a trip diary and to record their impressions. One student wrote:

> Before the EWB trip I assumed I would return to the US and look for an entry-level structural engineering job or an engineering consulting job … However, after being in Aceh I realized how much I enjoy humanitarian work. I want to find a way to combine engineering and humanitarian work, perhaps in international development work or disaster relief aid. If I can't find anything like that early in my career, then at the very least I plan to join a professional chapter of EWB to keep me balanced while I am working my boring engineering desk job.

David Schaad is just one of perhaps hundreds of engineering professors who work with Engineers Without Borders, sending students to remote parts of the world to help the less fortunate. The most important result of these projects is changing the world views of engineering students who discover that they want to continue using their engineering skills for peace engineering.

Camille George (1956–)

Fig. 7.12 Camille George

Work in developing countries can be frustrating. Often it seems that a lot of effort goes to helping just one or two people. Successful projects, however, can be replicated and end up helping thousands. One of these projects is run by Camille George, a mechanical engineering professor at the University of St. Thomas. She says, with great conviction: "My five-year goal is to feed 100,000 Haitian school kids a day."

George started her college career with a liberal arts degree from the University of Chicago and finished with a PhD in mechanical engineering from the University of Minnesota. Her efforts thus far have produced numerous labor-saving devices suitable for use in developing countries, including a manual shredding device, a labor-reducing mixer, a low-power cooling system, and a solar-powered water pasteurization system.

Each project has used engineering to empower impoverished women and enable them to profit from their countries' natural resources.

A member of Engineers Without Borders and Engineers for a Sustainable World, George led a project in which students helped women's cooperatives in

Haiti harvest breadfruit for use as a flour substitute. Another endeavor found George and her students in the West African country of Mali to investigate how to make the production of shea butter more efficient. (Shea butter is a natural fat extracted from the shea tree and is used as a cooking oil in West Africa.)

At the present time she has five students working on breadfruit processing. She has cooperation from over ten "laboratories" all over the world, including those in Haiti, the Marshall Islands, Columbia, Samoa, and the Philippines. She has engaged several systems engineers from the defense industry to figure out how to get the small manual system capable of processing one ton of breadfruit flour in a season out to the hundreds of communities that could benefit from this technology. She is working on a parallel process on the growing of seed potatoes in the Sahel, the semiarid tropical savanna ecoregion in Africa, which forms the transition between the Sahara desert and the savanna belt to the south. This region has always been concerned with food security.

George believes that engineering can help to build a more sustainable and just world. Speaking for many engineers, she says, "I get a lot of people who don't think this engineering is up to snuff, that it's kind of weak or lame. But I am going to prove them wrong because my engineering will impact the lives of people who have been traditionally neglected by mainstream engineering. I think this is true engineering, thinking outside the box to create simple solutions that can radically transform lives" [10].

Don O'Neal (1933–)

Fig. 7.13 Don O'Neal

After a solid Aggie education in mechanical engineering from Texas A&M, O'Neal went to work in the corporate world, including stints with Texas Instruments and as president of Applied Computer Products. He retired from corporate work in 1984 and wondered what to do next.

He decided to lend a hand south of the border, and his first project was to organize a team of medical doctors to go into remote areas of Guatemala to do surgical procedures. While there, he built several buildings, including a school in Santa Avelina. Living among natives in remote areas, he became aware of the danger caused by open fire cooking stoves used by the women. Often these stove exploded, causing severe burns. He figured he could make a better stove and started to tinker around, using suggestions of women in the village. He finally settled on a design that was both simple to use and inexpensive to construct. These stoves were named "onil" stoves by the women in the village, and the name stuck.

Fig. 7.14 A Guatamalan woman using an "onil" stove

O'Neal field-tested the stove in hundreds of homes, making modifications as needed, and finally put the stoves into production. The hardest part of the project was to convince the women to use them instead of cooking on open fires in their homes, the traditional way to cook. The women were quite aware that the open fires resulted in accidents, often resulting in severe burns. O'Neal also convinced them that when the huts were sealed tight, extremely high levels of carbon monoxide might occur when open fires were used. While the US EPA-recommended limit of indoor CO is less than 9 parts per million, O'Neal had been measuring CO concentrations exceeding 160 ppm in some huts. This level of CO interrupts the normal supply of oxygen to occupants and causes brain damage and loss of vital bodily functions. After O'Neal had installed his stoves, the CO levels in the huts rarely exceeded 5 ppm.

Convinced that his stoves were not only safer than the old means of cooking but that they needed only about one third of the wood the old stoves used, thus reducing the time women needed to gather firewood, O'Neal obtained funding from several charitable sources and went into production. He set up two factories to manufacture stoves in Guatemala and one in Mexico, and these are now producing stoves at a rate of 400 per month. There are over 60,000 "onil" stoves in Guatemala, Nicaragua, Honduras, and Mexico. O'Neal is presently submitting grant proposals to accelerate production of the stoves.

O'Neal, the epitome of a peace engineer, was recently honored with the Ashden Award in the UK, presented by Prince Charles.

Jordin Kare (1956–)

Fig. 7.15 Jordin Kare

Malaria has been eliminated in the USA but is still a dreaded disease in most tropical countries. Transmitted to humans by the *anopheles* mosquito, malaria kills over one million people annually. The most effective way to control the disease appears to be elimination of the vector, the mosquito. At present there are numerous efforts around the world focusing on the mosquito, including breeding malaria-parasite-free mosquitoes, knocking out the mosquito's sense of smell, and using mosquitoes as "flying syringes," carrying vaccine to unsuspecting people. One likely-to-succeed method of malaria control is zapping them with tiny lasers.

Jordin Kare and his colleague Lowell Wood have developed a device that can identify a mosquito from its shadow and its frequency of wing beats and then kill it with a low-energy laser. The method distinguishes between mosquitoes and other insects and can even distinguish a male mosquito from a female mosquito – an important distinction because it is only the female that stings humans and transmits the malaria parasite. The developers believe that this method can be used to kill billions of mosquitoes every night.

Kare took a circuitous path to his present work on malaria prevention. Following an electrical engineering degree from MIT, he went to the Lawrence Livermore National Laboratory to work on the Star Wars project, but left and now concentrates on "making a dent in a war that's gone on a lot longer [than the Cold War] and claimed a lot more lives" [11].

7.2 Peace Engineering at American Universities

It is my view that peace engineering has to start at the universities. Organizations such as Engineers Without Borders and Public Interest Research Groups (PIRGs) have reached the same conclusion and have actively promoted the formation of student chapters. In order for these chapters to be successful, however, they need to have the support of faculty, and in order for peace engineering to take root, it likewise needs the support of the faculty.

Most educators will not have an opportunity to practice peace engineering in the way Don O'Neal, Dennis Warner, and Ben Linder have done, but this does not mean that they are powerless to do anything, as David Schaad and Camille George illustrate. Engineering educators can participate in the peace engineering movement by reconsidering their role as faculty. In academics the three pillars of responsibility are service, teaching, and research (scholarly activity and publishing); peace engineering can play an important role in all of these.

7.2.1 Research and Scholarship

As noted in earlier chapters, military funds and military engineering research is pervasive on American campuses. This is no accident. Vannevar Bush, a founder of the National Science Foundation, was blunt about the main purpose of the new organization: "There must be more research and more adequate military research in peacetime … This can best be done through a civilian-controlled organization with close liaison with the Army and Navy …" [12]. It is difficult to estimate the amount of university research related to peace engineering, but it ought to be clear that this would, in terms of both funding and number of engineers, pale in comparison to military research.

The effect of military research at engineering schools has largely been ignored by the disciplines that study engineering and engineering education. Philosophers of engineering, those scholars who have devoted their professional careers to understanding the role of engineers in our society, seem oblivious to the greatest single decision engineers have to make – whether or not to work in the armaments industry. A recent conference of the Society of Philosophy and Technology, which has as its objective the understanding of the engineering profession, published the discussion topics for which it solicited learned papers. These were:

1. converging technologies and human enhancement;
2. converging technologies and engineering sciences;
3. converging technologies and risks;
4. converging technologies: general issues;
5. ethics and politics of emerging technologies;
6. philosophy and ethics of biomedical and nanotechnology;
7. philosophy and ethics of information technology;
8. environmental philosophy and sustainable technology;
9. philosophy of engineering and design;
10. robots, cyborgs, and artificial life;
11. technology and moral responsibility;
12. technology, culture, and globalization;
13. the good life and technology;
14. philosophy of technology: general and assorted issues;
15. reflective engineering.

Not one mention of war or the importance of the military in the engineering profession appears in this list.

Because of the overwhelming importance of military research in almost all engineering disciplines, most engineering scholars are unaware that it is possible to forge a successful academic career in engineering by doing peace-related research and scholarship instead of military work. Scholarly activity is mandatory for tenure and promotion in almost all universities; the choice of a personal research agenda is one of the most important decisions made by a young faculty member.

Consider the emerging field of synthetic biology. In 1989 scientists in Switzerland created DNA containing two artificial generic "letters" in addition to the four that appear naturally. This research has led other scientists to the possibility of building new life from interchangeable DNA parts. Such synthetic biological machines would work inside living cells, thus obtaining energy, nutrients, and other raw materials from cell metabolism. It is now feasible to assemble biological systems from interchangeable parts, just as machines are constructed from interchangeable parts, each of which has its own purpose.

Synthetic biology offers unparalleled opportunities for work beneficial to people and to the planet. For example, one great and tragic residual of regional conflicts, especially where there has been no battlefront separating armies, is antipersonnel land mines.

Land mines have two horrific attributes: they cannot be aimed and they do not decay. They cannot distinguish between a child and a soldier, and they can stay in the ground, unnoticed, for decades. When a child steps on a mine, the blast may kill, but more often it causes injuries requiring amputation. The number of people who have been maimed by land mines is staggering. In Cambodia alone there are over 35,000 amputees injured by land mines, while hundreds of thousands have died in the fields from loss of blood. Land mines are now a daily threat in Afghanistan, Angola, Bosnia, Cambodia, Chechnya, Croatia, Iraq, Mozambique, Nicaragua, Somalia, and dozens of other countries [13].

Land mines are designed and produced by engineers. The leading manufacturers and exporters of land mines are historically the USA, China, Italy, and the former Soviet Union. More than 50 countries have manufactured as many as 200 million antipersonnel landmines in the last 25 years. Some new developments in land mines are plastic mines that cannot be found with metal detectors, remotely scattered mines distributed from aircraft, and self-destructing mines that deactivate over a period of time. Although the use of self-destructing mines would eventually reduce the threat of land mines, these mines in their active phase still cannot discriminate between civilians and combatants.

There presently is no effective way to clear land mines. Sometimes metal detectors are used, sometime dogs are used to sniff out the TNT, and sometimes metal prodders can be used to probe the ground. Demining technology is not nearly as advanced as the construction and deployment of new landmines, and the problem seems to be accelerating as regional conflicts continue.

However, new ideas are being developed for detecting and dismantling land mines, and one of these is the use of synthetic biology. Using interchangeable DNA parts, engineers can now modify organisms to have them do what is specifically needed. One application of synthetic biology, presently in the development phase, is the modification of metabolic circuits so that if the plant detects the presence of TNT it will glow red, and at lower levels of TNT it would have a yellow color, while in the absence of TNT the plant would be green. By spreading these genetically altered plants over a suspected mine field the red dots with yellow circles would be the bull's eyes showing the presence of a land mine [14].

One point here is that technology that helps people can be just as "cool" as technology that kills people. The trick is to be clever enough as a faculty scholar to find these projects. Rober Fulgrum, a Unitarian minister and bestselling author, has a suggestion:

> Maybe we should develop a Crayola bomb as our next secret weapon. A happiness weapon. A beauty bomb. And every time a crisis developed, we would launch one. It would explode high in the air – explode softly – and send thousands, millions, of little parachutes into the air. Floating down to earth – boxes of Crayolas. And we wouldn't go cheap, either – not little boxes of eight. Boxes of sixty-four, with the sharpener built right in. With silver and gold and copper, magenta and peach and lime, amber and umber and all the rest. And people would smile and get a little funny look on their faces and color the world with imagination [15].

A problem faced by faculty at research universities is the question of funding. Funding from the military is fairly easy to get and the work is often interesting. But restrictions on this funding can make the military a less than optimum source. An alternative source is industry. Industry funding, however, also comes with strings attached. Researchers and their students have to sign confidentiality agreements and have to agree to seek approval for all publications. From industry's perspective, this is only reasonable, because the purpose of the funding is to enhance profits for a given company. From an academic standpoint, these restrictions are often untenable. There are many ethical considerations involved in accepting research money from industry, including the misrepresentation and withholding of facts that might cause damage to the industry [16]. The objective of academicians who want to do peace engineering research is to find industries that can see future profits in this new research.

7.2.2 *Service*

Faculty members are expected to perform service to both the university and society. Land grant colleges generally expect a higher commitment to such outreach, but all universities look with great favor on faculty who are actively involved in university life as well as community life.

One way engineering faculty can serve the university is to advise and sponsor student organizations. An example of such an organization is Engineers Without Borders (EWB), an organization modeled after Doctors Without Borders. EWB now has 320 chapters at almost all major engineering schools; most chapters were started and are run by engineering students. Advising one of these chapters is a positive and effective way for faculty to practice peace engineering. Another example is Public Interest Research Group (PIRG), which has hundreds of chapters on college campuses. Issues addressed by US PIRG, the parent organization, include winning protection for wilderness preserves, advocating the strengthening of hazardous waste legislation, supporting funding for renewable energy, agitating for openness in consumer contact with hazardous materials, backing bans on drilling in national monument lands, and helping pass stricter standards on drinking water. PIRG chapters on university campuses select their own local projects while working to support the national objectives. The objectives of PIRGs have been mostly limited to environmental matters. Finally, university faculty might consider launching student groups whose primary objective is preparation for careers in peace engineering.

7.2.3 Teaching

Engineering educators have meager resources for making their students aware of career options that do not involve military engineering. The fact that so many engineers work for the military and the moral conflicts that this might present is seldom if ever mentioned in engineering teaching material. Papadopoulos surveyed the Engineering Case Library, which contains 261 cases, and, using a keyword search, found that only 15 of the cases had any mention of military or defense or armaments. Only 6 of the 15 dealt in any depth with ethical issues of working for the military. A review of textbooks on engineering ethics found that of 17 books surveyed, only 7 included any mention of military or defense-related ethics issues [17]. Although military engineering permeates all engineering research, it is almost invisible in undergraduate engineering education.

Asking moral questions relating to military engineering may be considered by some to be indoctrination. There is a difference between teaching and indoctrination, however, and the latter is anathema to education because it takes away the fundamental attribute of an educated person – the ability to think critically. Professors who indoctrinate (and there are some) are perhaps the most dangerous of all teachers because they pretend to promote clear thinking, but all the while press their own views on students and try to convince them that the professor's views are actually products of rational thought.

The fact that teachers help their students to think rationally and critically has made this profession dangerous to all totalitarian states. In 1940 when the Soviet Union invaded the Baltic countries of Estonia, Latvia, and Lithuania, for example, the Soviets systematically deported the most dangerous people to Siberia, and the first ones on the cattle trains were the teachers and professors. With the real teachers gone, the Soviets placed their own "teachers" in the schools – teachers who could, through indoctrination, be relied on to press communist values and a communist economic agenda.

Such tragedies seldom occur in a free country, but this does not reduce our need for vigilance. The only defense against indoctrination is clear thinking in place of propaganda, and it is the responsibility of academicians to make sure this occurs in education, including engineering education.

Consider a session I once taught in a professional ethics course. In this course we tried hard to take issues apart and to discover what values drive decisions, and how a difference in values can lead to significant disagreements. The best examples to use in such a course come from unpredictable sources. For example, we had received a campuswide notice to come to a rally in support of our soldiers who had been fighting overseas, and some of the students wanted to talk about it. We decided that when one goes to such a rally (perhaps carrying an American flag?), one is sending mixed signals. What exactly is being supported?

The students decided that there are three different recipients of such a show of support:

1. America as a nation and as an idea,
2. the soldiers who are placed in harm's way,
3. the political leaders who place the soldiers in harm's way.

None of us had any problems with supporting the first two, but we could not figure out how to show support for the first two without also unwittingly showing support for the last one. I did not tell them to avoid the rally, and I would never offer my own reason for not going, but the discussion allowed them to think through the problem.

Students are often unaware of the choice they have to make when they graduate with an engineering degree. They can choose to work in defense-related industry, or they can choose an alternative path. It is our role as educators to make this choice clear to them.

Rosemary Chalk is right on the money when she makes this recommendation:

> In the education of young scientists and engineers, historical and critical examinations of the relationships between science and the military should be encouraged. Students should be informed early in their training of some of the ethical dilemmas raised in the course of military research. They should be asked to consider the likelihood that they will be asked to work on projects that are not consistent with their own moral standards, and that should be provided with examples illustrating alternative approaches to ethical dilemmas in military research [18].

The first attempt to demonstrate to students that they will have to make moral decisions about where they will work and what kind of engineering career they will have was a conference held at Bucknell University in 2004. The book, based on papers delivered at the conference, is entitled *Peace Engineering: When Personal Values and Engineering Careers Converge* [19].

Just as there are colleges for the military, there are those that have embraced the concept of peace and developed their programs to study and teach about peace. The United Nations, recognizing that an understanding of peace is vital to the world, established a university dedicated to the scholarly pursuit of peace. The University for Peace is in Costa Rica and has as its mission "to provide humanity with an international institution of higher education for peace and with the aim of promoting among all human beings the spirit of understanding, tolerance and peaceful coexistence." The campus has about 170 students from over 50 countries and grants master;s and doctoral degrees [20]. The United States Association for the University for Peace is an organization that supports the university.

We don't need a whole university to teach peace engineering, however, and peace engineering does not even have to be taught in ethics courses. The responsibility of engineers to the public is an overarching principle, and its importance can be demonstrated in almost any course. I will close with just one example of how this can be done.

I used to teach a freshman environmental engineering course that was designed to draw people into the civil engineering department. Since this was not a required course within the curriculum, I had some leeway in how I taught the course. Along the way we had numerous discussions about engineering and professional responsibility. At one point I assigned them a homework problem. Here it is, verbatim:

> According to the statement of the Draeger Works in Luebeck, in the gassing of the whole population in a city, only 50% of the evaporated poison gas is effective. The atmosphere must be poisoned up to a height of 20 meters at a concentration of 45 mg/m^3. How much phosgene is needed to poison a city of 50,000 inhabitants who live in an area of four square kilometers?

The problem comes from a German high school textbook, published in the late 1930s. It was passed on to me by a colleague at Bates College.

I assigned the problem without comment. The day the problem set was due I collected the problems and then, as usual, asked if anyone had any difficult with them. On this day, there invariably would be one or two people who would say something like, "I didn't do the gassing problem."

"Why?" I would ask.

"I just don't want to do problems like that," the student would respond.

What we then had was a teachable moment. Why should an engineer not do some kinds of problems? What kind of problems are these, and why should they not be solved? This is education at its best.

7.3 Conclusion

Engineers can and do apply their technical skills in an almost infinite number of ways. Some of these are directly in the research, design, and production of military hardware. But even these military engineers would agree, I believe, that the world would be a better place if military engineering became unnecessary. The ultimate goal of peace engineering – the use of engineering skills for the promotion of peace – is exactly that: to make military engineering obsolete and unnecessary. This is a morally admirable goal, worth considering by young people who are just starting out in their engineering careers.

References

[1] Agency for Toxic Substances and Disease Registry (2009) US Department of Health and Human Services, Atlanta, GA
[2] Rather D (1987) CBS News. 29 April 1987
[3] Soomere T (2005) Scandinavian Shipping Gazette. 25 February 2005
[4] Home-Douglas P (2004) The water guy. Prism Mag 13(3): http://findarticles.com/p/articles/mi_qa3797/is_200411/ai_n9460395/
[5] Anderson S (1995) The man who tried to save the world: the dangerous life and mysterious disappearance of an American hero. Doubleday, New York
[6] Shawcross W (1995) A hero of our time. New York Review of Books, 30 November 1995
[7] A laser inventor ending arms work (1986) New York Times, September 11, Section A, p 18
[8] World Water Council, http://www.worldwatercouncil.org (last access Nov. 9 2009)
[9] Department of Environmental Science and Engineering, School of Public Health (2007) ESE Notes. University of North Carolina, Chapel Hill, NC
[10] University of St. Thomas, http://www.stthomas.edu (last access Nov. 9 2009)
[11] Guth RA (2009) Rocket scientists shoot down mosquitoes with lasers. Wall Street Journal, 14 March 2009
[12] Bush V (1945) Science: the endless frontier. US Government Printing Office, Washington, DC
[13] International Campaign to Ban Landmines, www.icbl.org (last access Nov. 9 2009)
[14] Looger LL, Sawyer MA, Smith JJ, Hellinga HW (2003) Computational design of receptor and sensor proteins with novel functions. Nature 423185-190, 8 May
[15] Fulgrum R (1990) All I really need to know I learned in kindergarten: uncommon thoughts on common things. Villard Books, New York
[16] Schrag B, Ferrell G, Weil V, Fiedler T (2003) Barking up the wrong tree? Industry funding of academic research. A case study with commentaries. Sci Eng Ethics 9(4) pp. 569–582
[17] Papadopoulos C, Hable A (2008) Including questions of military and defense technology in engineering ethics education. In: Proceedings of the American Society for Engineering Education Conference and Exposition, Washington, DC
[18] Chalk R (1989) Drawing the line: an examination of conscientious objection in science. In: Mitcham C, Siekevitz P (eds) Ethical issues associated with scientific and technological research for the military. New York Academy of Sciences, New York
[19] Vesilind PA (ed) (2006) Peace engineering: when personal values and engineering careers converge. Lakeshore, Woodsville, NH
[20] United Nations General Assembly (1980) Resolution Session 35, Establishment of the University for Peace. 5 December 1980

About the Author

Following his undergraduate degree in civil engineering from Lehigh University, Vesilind received his PhD in environmental engineering from the University of North Carolina in 1968. He spent a postdoctoral year with the Norwegian Institute for Water Research in Oslo and a year as a research engineer with Bird Machine Company. In 1970, he joined the faculty at Duke University, where he served as chair of the Department of Civil and Environmental Engineering. In 1999 he was appointed to the R. L. Rooke Chair of the Historical and Societal Context of Engineering at Bucknell University. He served in this capacity until his retirement in 2006.

In 1976–1977 he was a Fulbright Fellow at the University of Waikato, Hamilton, New Zealand. He's a former trustee of the American Academy of Environmental Engineers, a past president of the Association of Environmental Engineering Professors, a Fellow of the American Society of Civil Engineers, and a registered professional engineer in North Carolina. He is the recipient of the 1970 Collingwood Prize, awarded by the American Society of Civil Engineers. Other awards include the E. I. Brown Award from the students of the Department of Civil Engineering, Duke University, for teaching excellence (four times), the Tau Beta Pi Teaching Award from the students of the School of Engineering, Duke University (1999), the 2005 Award for Achievement in Environmental Engineering Education, American Society of Civil Engineers, and the Distinguished Service Award from the Association of Environmental Engineering and Science Professors, 1998 and 2005.

While at Duke he headed for many years the Program in Science, Technology and Human Values, an undergraduate enrichment program that sought to build bridges between the humanities and engineering.

His research has resulted in the authorship of over 174 articles in professional journals and 28 books on environmental engineering and professional ethics, and he has been the primary adviser to 12 PhD students and 46 master's students. He has also been a principal investigator on over 53 research projects and training grants.

Index

MIX
Papier aus verantwortungsvollen Quellen
Paper from responsible sources
FSC® C105338

If you have any concerns about our products,
you can contact us on
ProductSafety@springernature.com

In case Publisher is established outside the EU,
the EU authorized representative is:
Springer Nature Customer Service Center GmbH
Europaplatz 3, 69115 Heidelberg, Germany

Printed by Libri Plureos GmbH
in Hamburg, Germany